Psychological Themes in the School of Alexius Meinong

Meinong Studies / Meinong Studien

Edited for / Herausgegeben für

Alexius-Meinong-Institut –
Forschungsstelle und Dokumentationszentrum für
Österreichische Philosophie, Karl-Franzens-Universität Graz

by / von Mauro Antonelli, Marian David

Volume / Band 10

Psychological Themes in the School of Alexius Meinong

Edited by
Arnaud Dewalque and Venanzio Raspa

DE GRUYTER

ISBN 978-3-11-066251-1
e-ISBN (PDF) 978-3-11-066485-0
e-ISBN (EPUB) 978-3-11-066265-8
ISSN 2198-2309

Library of Congress Control Number: 2019952791

Bibliographic information published by the Deutsche Nationalbibliothek
The Deutsche Nationalbibliothek lists this publication in the Deutsche Nationalbibliografie; detailed bibliographic data are available on the Internet at http://dnb.dnb.de.

Printing and binding: CPI books GmbH, Leck

www.degruyter.com

Contents

Arnaud Dewalque and Venanzio Raspa
Introduction — 1

Riccardo Martinelli
Meinongian Psychology — 11

Denis Seron
Color Geometry – Or Color Grammar? — 33

Carlo Ierna
Thinking the Impossible: The *Gestalt* of a Round Square — 47

Philipp Berghofer
What Role Can Experimental Psychology Play in Philosophical Investigations: Meinong and Benussi as Precursors of Experimental Philosophy? — 61

Arnaud Dewalque
A Critique of Meinongian Assumptions — 85

Cristina Travanini
Translating *Über Annahmen*: Questions and Perspectives — 109

Gemmo Iocco
Assumptions and Objective Value-Theory: Meinong's Psychology of Valuation — 125

Venanzio Raspa
Meinong on Emotional Truth — 145

Erwin Tegtmeier
Meinong's Realist Analysis of Perception — 161

Mauro Antonelli
Vittorio Benussi's Genetic-Experimental Phenomenology of Perception and its Place in the History of Gestalt Psychology — 169

Index of Names — 199

Arnaud Dewalque and Venanzio Raspa

Introduction

The contributions gathered in this volume address various facets of the philosophical psychology elaborated on by Alexius Meinong and some of his students. They cover a wide range of topics, from the place of psychological investigations in Meinong's philosophical programme to his thought-provoking views on perception, colours, *Vorstellungsproduktion*, assumptions, values, truth, and emotions. Although psychological themes are omnipresent in the works of Meinong and his students, it is probably not unfair to say that their theory of the mind received considerably less attention in literature than their object theory, which somehow became the hallmark of the Meinong School. Our hope is that this volume will help restore the balance and, indirectly, foster a novel understanding of Meinong's philosophy in the context of his time.

Meinong's Philosophy in Context

Understanding Meinong's philosophy in context amounts to seeing it as the result of a development which is both internal and external to Meinong's thought, and therefore as the result of a broad intellectual process. It means explaining how his philosophical views developed from, and interacted with, other, competing views. In this respect, it is important to recall that Meinong's philosophy – his object theory – is the upshot of collective work carried out by Meinong himself and some of his students. It is common knowledge that Witasek and Benussi assisted Meinong in his psychological investigations, while Mally and Ameseder contributed to the development of the object theory. Furthermore, this collective work also bears the mark of philosophers who indirectly contributed to the development of Meinongian philosophy (like Bolzano, Brentano, Twardowski) or debated with Meinong and his heirs (Husserl, Russell, Lipps), thus pushing the former towards a refinement of his views.

Over the last decades, a great amount of attention has been devoted to the object theory. Accordingly, the primary focus of Meinong studies so far has been

Arnaud Dewalque, University of Liège / University of Luxembourg
a.dewalque@uliege.be
Venanzio Raspa, Università degli Studi di Urbino Carlo Bo
venanzio.raspa@uniurb.it

https://doi.org/10.1515/9783110664850-001

on logic, semantics, and formal ontology, as manifest in the writings of Héctor-Neri Castañeda, Karel Lambert, Richard Routley (Sylvan), Terence Parsons, Edward Zalta, William J. Rapaport, Jaakko Hintikka, Dale Jacquette, and Jacek Paśniczek – to name but a few. This line of research has great merits. In particular, it contributed to develop Meinong's insights into directions that he himself could not have predicted. Moreover, the expansion and modification of his views in line with the results obtained on the basis of scientific investigations is entirely consistent with Meinong's scientific spirit, especially if it can help address current philosophical issues. While such studies drawn on Meinong's writings to improve some of his most interesting theses (as Parsons, Lambert, Jacquette, and Paśniczek themselves state), some related debates – like those opposing Neo-Quinean and Neo-Meinongian philosophers – developed more independently from the study of Meinong.

However legitimate, such lines of inquiry are not the only ones which are worth pursuing today. Our own feeling is that addressing Meinong's philosophy in context does not simply mean to do some philology. We can start with the reading of the writings by Meinong and his students, and then move on to examine current issues, or put the philosophy of the Graz school in relation with other philosophers – and, of course, criticise it. All these approaches amount to revitalising Meinong's fundamental intuitions in the present-day research situation while accepting what he calls the principle of critical non-conclusiveness of knowledge.

Now, if one looks back at Meinong's writings in their historical context, the importance of his psychological investigations could hardly be overestimated. Like Brentano before him, Meinong takes it that (scientific) philosophy is not possible without (descriptive) psychology. This is not to say that philosophy and psychology are one and the same discipline. Rather, psychology is but a "part" of philosophy, indeed its "fundamental discipline", while philosophy is the name of "a whole group of sciences", whose commonality is that they are all dealing with mental phenomena in some way.[1] On Meinong's view, thus, there is a pretty tight connection between philosophy and psychology. Although he most decidedly rejected any form of psychologism, he never gave up the thought that philosophy was not separable from psychological investigations. In this respect, his reaction to the antipsychologist struggle is quite telling. When, in 1912, the Neo-Kantian philosopher Heinrich Rickert asked him to sign a joint statement against the attribution of philosophical positions to experimental

1 See Meinong 1885, *GA* V, p. 5.

psychologists, Meinong replied by highlighting again the importance of psychology for philosophy: "As you know, my personal field of work grew ever more apsychological. And yet, it seems to me inconceivable to forget the endeavours which arose for me from the close contact with the lively psychological empirie [...]. I wouldn't like to encourage the tendency to separate experimental psychology, and thereby psychology in general, from its connection with the remaining philosophical disciplines".[2]

Meinongian Psychology

Meinong's ten "Essays on Psychology" (*Abhandlungen zur Psychologie*) gathered by Alois Höfler in 1913 within the framework of the *Collected Essays* (*Gesammelte Abhandlungen*) were intended as contributions to the then ongoing psychological research.[3] The same holds true of his two important books *On Assumptions* (*Über Annahmen*) and *On Emotional Presentation* (*Über emotionale Präsentation*). As is obvious from those writings, Meinong's views about mental phenomena are not a mere repetition of Brentano's, in fact they sometimes present themselves as corrections of things Brentano said. This raises the following question: What, if any, are the distinctive features of Meinongian philosophical psychology? Putting aside a number of analyses dedicated to more or less local phenomena, we believe that Meinongian psychology as a whole is characterised by a *triad of distinctions*, namely: the distinction between (1) content and object, (2) activity and passivity, (3) mental phenomena and mental dispositions. Let us briefly comment on each distinction in turn.

1. The distinction between content and object is usually traced back to Twardowski's habilitation thesis *On the Content and Object of Presentations* (*Zur Lehre vom Inhalt und Gegenstand der Vorstellungen*). Yet, Twardowski acknowledges that he himself took it from the Höfler-Meinong handbook of logic:

> What we called 'content of the representation [*Vorstellung*] and the judgement' lies just as much completely within the subject as the act of representation and of judgement itself. The words 'thing' and 'object' are used in two senses: on the one hand for that *independently existing* entity [...] at which our representation and judgement aim, as it were; on the other

---

2 Undated letter to Rickert (reply to Rickert's letter of the 7th December 1912), Meinong-Nachlaß, Box LX, n° 6224, quoted in Dölling 2001, p. 157.

3 See Meinong 1921, *GA* VII, p. 9.

> hand, for the mental, more or less approximate, 'picture' of that real entity which *exists 'in' us*.[4]

By systematically applying the content-object distinction to both presentations and judgements, Twardowski moves away from Brentano's traditional doctrine. However, he still remains close to Brentano insofar as he maintains that representations and judgements refer to the same object, that is, to the representational object (*Vorstellungsgegenstand*).[5] In a letter to Meinong dated 11th July 1897, Twardowski speaks of the state of affairs (*Sachverhalt*) as the object of judgement.[6] In his reply, Meinong states that his epistemological viewpoint is not fully expressed in the *Logik* he wrote with Höfler.[7] In his review of Hillebrand's book on inference (*Die neuen Theorien der kategorischen Schlüsse*), he introduced a new class of judgements, namely: relational judgements (*Beziehungsurteile*), which are a first step towards the postulation of judgements which have a state of affairs as their object. Twardowski himself made a similar move in his 1894/95 logic lessons.[8] The thoughts of Twardowski and Meinong clearly exhibit some affinities in respect to the distinction between content and object of judgements and (re)presentations. In the essay "On objects of higher-order" ("Über Gegenstände höherer Ordnung und deren Verhältnis zur inneren Wahrnehmung"), although Meinong refers to Twardowski's habilitation thesis only in a footnote, he nevertheless takes up the latter's arguments in favour of this distinction, as is obvious from a comparison between the two texts.[9] More importantly, Meinong presents the distinction of content and object – which originates from considerations "in part really, and in part only supposedly, psychological"[10] – as a crucial step towards the object theory. Indeed, the distinct types of objects (objecta and objectives) correspond to the content of representations and judgements, respectively. Meinong dealt with this view in several writings throughout his career. In his *On Emotional Presentation*, the notion of content is defined as that part of an experience (*Erlebnis*) which varies or remains constant regardless of the variations of the cor

4 Höfler 1890, § 6; Twardowski 1894, p. 4 [1977, p. 2].
5 See Twardowski 1894, § 4, p. 9, § 7, p. 38.
6 See Meinong & Twardowski 2016, p. 85.
7 See Meinong & Twardowski 2016, p. 92.
8 See Twardowski 2016, p. 34–35, 91–92.
9 See Twardowski, 1894, § 6, p. 30–34; Meinong 1899, *GA* II, p. 186–188.
10 Meinong 1904, *GA* II, p. 503 [1960, p. 94].

responding object.[11] It is a central tenet of Meinong's psychological theory of presentation (*Präsentation*).

2. One further distinctive feature of Meinong's theory of the mind lies in the claim that the mind is *active* and *productive*, or that many mental phenomena are best described as involving, as Höfler puts it, some "mental work".[12] Interestingly, the activity-passivity distinction seems to have been utterly absent from Brentano's *Psychology from an Empirical Standpoint*. The notion of "mental action" (*psychische Aktion*) is, however, clearly stressed by Meinong in his 1894 "Contributions to the Theory of Psychological Analysis" ("Beiträge zur Theorie der psychischen Analyse"). Distancing himself from Carl Stumpf, he writes:

> The psychological analogon of the opposition between movement and rest is provided by the opposition between activity and passivity, which in [the psychological] field is no less fundamental than the former in the physical field.[13]

Arguably, this view had significant consequences for Meinong's way of conceiving of mental phenomena. First of all, it led him to endorse a quadripartite classification of mental phenomena into (i) representations, (ii) thoughts, (iii) feelings, and (iv) desires. This division may be arrived at by adding the activity-passivity distinction to Aristotle's distinction between "cognition" (*noûs*) and "desire" (*orexis*): representations are passive cognitive phenomena, while thoughts are active cognitive phenomena; similarly, feelings are passive emotional phenomena, while desires are active emotional phenomena. Another consequence of this view is the famous introduction of "assumptions" as an "intermediary field" between representations and judgements. Very roughly, Meinongian assumptions are cognitive activities of the mind which lack the belief component proper to judgement.

3. Probably no less central to Meinong's philosophical psychology is the distinction between mental *phenomena* and mental *dispositions*. It is common knowledge that Herbart rejected the so-called "psychology of faculties" or "psychology of powers" (*Vermögenspsychologie*) as unscientific. Yet, Meinong notices, it is striking to see how the thought of a capacity, or power, still

---

11 See Meinong 1917, *GA* III, p. 339, 347 [1972, p. 49, 55].
12 Höfler 1894.
13 Meinong 1894, *GA* I, p. 382. See also Höfler 1894, p. 31; 1930, p. 106, fn. 1.

"permeates the most ordinary conception of mental life".[14] The fact is, there are many aspects of mental life which seem unaccountable without having a full-blown theory of dispositions in hands. Meinong's interest in dispositions dates back at least to the beginning of his academic career. When he took up his professorship at the University of Graz in 1882, he held a course on "Educational Theory, or Psychology of Dispositions" (*Pädagogik oder Psychologie der Dispositionen*).[15] For more than two decades, he then worked on a theory of dispositions which he made public in his lecture courses only,[16] before eventually publishing a sketch thereof in the *Festschrift* for Eduard Martinak's 60th anniversary.[17] One idea that became prominent in the Meinong School was that the goal of teaching is to create in the learner some "dispositions to judge in an evident way".[18] Yet, considerations on dispositions are far from being limited to education. In 1889 Meinong insisted that the theory of dispositions is a prerequisite for a correct understanding of imagination (*Phantasie*).[19] Similarly, in 1894, Höfler argued for a dispositional theory of attention, according to which "attending to something" is best understood in terms of "making oneself ready for some mental work",[20] etc. Admittedly, the study of mental dispositions is not an independent chapter of psychology and cannot be entirely disconnected from the study of mental phenomena. After all, as Meinong puts it, "every disposition is specified, first and foremost, according to that which it is disposed to", that is, according to its phenomenal "correlate".[21] Still, Meinong argues, it is important to see that there is more to one's mental life than just mental phenomena.

14 Meinong 1889, *GA* I, p. 196.
15 See Höfler 1919, p. 24 fn.; 1921, p. 370.
16 According to Dölling 1999, p. 235–37, Meinong held a course *on Psychologische Prinzipien der Pädagogik* in the Summer Semester 1884, on *Psychologische Prinzipien der Pädagogik (Lehre von den psychischen Dispositionen)* in the Summer Semester 1887, and on *Psychologische Prinzipien der Pädagogik (Dispositionspsychologie)* in the Summer Semester 1892.
17 See Meinong 1919.
18 Höfler 1919, p. 24 fn.
19 See Meinong 1889, *GA* I, p. 196.
20 Höfler 1894, p. 100.
21 Meinong 1889, *GA* I, p. 197.

Plan of the Book

This volume comprises three sections. Section 1 is dedicated to the relationship between psychology and philosophy. It discusses critically the place of psychology in Meinong's programme, his theory of colours, and his theory of *Vorstellungsproduktion*. Section 2 is dedicated to Meinong's views on assumptions and emotions. Section 3 addresses Meinong's and Benussi's analyses of perception. It is plain that the contributions gathered in those three sections are very far from giving an exhaustive picture of Meinongian psychology. Much more needs to be said, especially about Meinong's theory of emotion, imagination and disposition. Yet, the present volume will have reached its goal if it gives the reader a taste of Meinong's theory of the mind and paves the way to a more thorough exploration and reception of the latter.

Acknowledgements

This special issue of the *Meinong Studies* is the outcome of a two-session conference held at the University of Liège (26–27 May 2016) and the University of Urbino (10–11 November 2016) under the title *Psychological Themes in the School of Alexius Meinong*. We are grateful to all the participants for their thought-provoking talks and to the authors who agreed to revise their paper for the present volume. We also wish to express our warmest thanks to the editorial board of the *Meinong Studies*, and especially to Mauro Antonelli for his friendly support and to Ulf Höfer for his patience and help with the preparation of the manuscript.

References

Dölling, Evelyn (1999), *"Wahrheit suchen und Wahrheit bekennen". Alexius Meinong: Skizze seines Lebens*, Amsterdam / Atlanta, GA: Rodopi.

Dölling, Evelyn (2001), "Alexius Meinong. Von der philosophischen Societät zum philosophischen Seminar", in: T. Binder *et al.* (eds.), *Bausteine zu einer Geschichte der Philosophie an der Universität Graz*, Amsterdam: Rodopi, p. 149–172.

Höfler, Alois (1890), *Logik. Unter Mitwirkung von A. Meinong*, Wien: Hölder.

Höfler, Alois (1894), *Psychische Arbeit*, Hamburg / Leipzig: Voss.

Höfler, Alois (1919), "Martinak als Pädagog und Pädagogiker. Praxis und Theorie in der Pädagogik", in: *Beiträge zur Pädagogik und Dispositionstheorie. Eduard Martinak zur Feier seines 60. Geburtstages dargebracht von Fachgenossen, Schülern und Freunden*, hrsg. von A. Meinong, Prag / Wien / Leipzig: Schulwissenschaftlicher Verlag A. Haase, p. 18–28.

Höfler, Alois (1921), "Meinongs Psychologie", in: *Zeitschrift für Psychologie und Physiologie der Sinnesorgane*. I. Abteilung: *Zeitschrift für Psychologie*, 86, p. 368–374.
Höfler, Alois (1930), *Psychologie*, 2nd ed. Vol. 1. Wien – Leipzig: Hölder / Pichler / Tempsky.
Meinong, Alexius (1885), *Über philosophische Wissenschaft und ihre Propädeutik*, Wien: Hölder; repr. in *GA* V, p. 1–196.
Meinong, Alexius (1889) "Phantasie-Vorstellung und Phantasie", in: *Zeitschrift für Philosophie und philosophische Kritik*, 95, p. 161–244; repr. in *GA* I, p. 193–271.
Meinong, Alexius (1894), "Beiträge zur Theorie der psychischen Analyse", in: *Zeitschrift für Psychologie und Physiologie der Sinnesorgane*, 6, p. 340–385, 417–455; repr. in *GA* I, p. 305–388.
Meinong, Alexius (1899), "Über Gegenstände höherer Ordnung und deren Verhältnis zur inneren Wahrnehmung", in: *Zeitschrift für Psychologie und Physiologie der Sinnesorgane*, 21, p. 182–272; repr. in *GA* II, p. 377–471.
Meinong, Alexius (1904), "Über Gegenstandstheorie", in: *Untersuchungen zur Gegenstandstheorie und Psychologie*, hrsg. von A. Meinong, Leipzig: J. A. Barth, p. 1–50; repr. in *GA* II, p. 481–530 [Engl. transl.: Meinong (1960)].
Meinong, Alexius (1917), *Über emotionale Präsentation*, in: *Kaiserliche Akademie der Wissenschaften in Wien*. Phil.-hist. Klasse, Sitzungsberichte, CLXXXIII, 2. Abh.; repr. in *GA* III, p. 283–476 [Engl. transl.: Meinong (1972)].
Meinong, Alexius (1919), "Allgemeines zur Lehre von den Dispositionen", in: A. Meinong (ed.), *Beiträge zur Pädagogik und Dispositionstheorie. Eduard Martinak zur Feier seines 60. Geburtstages dargebracht von Fachgenossen, Schülern und Freunden*, Prag / Wien / Leipzig: Schulwissenschaftlicher Verlag A. Haase, p. 33–54; repr. in *GA* VII, p. 289–310.
Meinong, Alexius (1921), "A. Meinong [Selbstdarstellung]", in: *Die deutsche Philosophie der Gegenwart in Selbstdarstellungen*, hrsg. von R. Schmidt, Bd. 1, Leipzig: Meiner, p. 91–150; repr. in *GA* VII, p. 1–62.
Meinong, Alexius (1968–78), *Alexius Meinong Gesamtausgabe* (*GA*), hrsg. von Rudolf Haller und Rudolf Kindinger gemeinsam mit Roderick Chisholm, 7 vols., Graz: Akademische Druck- u. Verlangsanstalt.
Meinong, Alexius (1960), "The Theory of Objects", translated by I. Levi, D. B. Terrell, and R. M. Chisholm, in: *Realism and the Background of Phenomenology*, ed. by R. M. Chisholm, Glencoe (Ill.): Free Press, p. 76–117.
Meinong, Alexius (1972), *On Emotional Presentation*, translated, with an introduction by M.-L. Schubert Kalsi, with a foreword by J. N. Findlay, Evanston (Ill.): Northwestern University Press.
Meinong, Alexius and Twardowski, Kazimierz (2016), *Der Briefwechsel*, hrsg. und eingeleitet von V. Raspa, Berlin / Boston: De Gruyter (= *Meinong Studies / Meinong Studien*, vol. 7).
Twardowski, Kazimierz (1894), *Zur Lehre vom Inhalt und Gegenstand der Vorstellungen*, Wien: Hölder. Unveränderter Nachdruck der ersten Auflage, Wien 1894. Einführung von R. Haller, mit einem Personen- und Sachregister von R. Fabian, München / Wien: Philosophia Verlag, 1982.
Twardowski, Kazimierz (1977), *On the Content and Object of Presentations*, Engl. transl. by R. Grossmann, The Hague: Nijhoff.

Twardowski, Kazimierz (2016), *Logik. Wiener Logikkolleg 1894/95*, hrsg. von A. Betti und V. Raspa, Berlin / Boston: De Gruyter.

Riccardo Martinelli

Meinongian Psychology

Abstract: Meinong's early writings include a number of discussions that are in line with the mainstream psychological research of his time. This is not at odds with the parallel theorizing tendency of his philosophical psychology. Even after developing the theory of objects, and despite handing psychological research over to his pupils, Meinong kept thinking highly of psychology. Meinong's psychology is in tune with his most relevant philosophical doctrines. The theory of "production" virtuously integrates that of founded objects. The manifold "dispositions" (including phantasy) and the different subjective "attitudes" (gestalt-like or analytical) lead to regular, lawful deviations from standard responses that can and should be investigated experimentally. While being an independent discipline, Meinongian psychology nevertheless fits into a full-fledged system of philosophical assumptions.

1 Critical Issues

To many philosophers, the theory of objects may well be the only relevant thing in Meinong's thought. Of course this is a legitimate view: in the long wake of Russell's criticism, Meinong's thoughts concerning ideal and non-existing objects are certainly among his most discussed contributions. Still, there is no point in making the theory of objects a sort of black hole, which swallows everything else that Meinong has done. From a historical point of view, this is simply wrong. In fact, such an interpretation is not only at odds with many of Meinong's own pronouncements, but it also prevents us from making sense of the relation between Meinong's thoughts on psychology and the activities he carried out at the Graz laboratory with his pupils. As far as psychology is concerned, Meinong has devoted much effort to both theoretical and empirical aspects of it, occasionally complementing his research with experimental studies.

In 1913, seven years before Meinong's death, Alois Höfler started collecting his writings: *Alexius Meinongs Gesammelte Abhandlungen* (Collected Essays). Of the three planned volumes, only the first two were eventually published: respectively *Abhandlungen zur Psychologie* (Essays on Psychology), and *Abhandlungen*

Riccardo Martinelli, Dipartimento di studi umanistici, Università di Trieste
E-mail: martinel@units.it

https://doi.org/10.1515/9783110664850-002

zur Erkenntnistheorie und Gegenstandstheorie (Essays on the Theory of Knowledge and the Theory of Objects).[1] In the Preface, reprinted in both volumes, Höfler explained why the collection was needed. Meinong had been frequently asked for a new edition of his barely available early articles, but was permanently too pressed by new projects. Thus, his pupils undertook the publication in his place, supplementing the essays with additional notes (*Zusätze*) in order to update them to "the current state" of 1913.[2]

Admittedly, Höfler had a hard time singling out Meinong's "psychological" essays from the rest of his production. He opens his introduction with the frank confession that the "ten psychological essays in this volume do not form a neat unity and do not exhibit a continuous, linear progress, as happens with the five essays of the second volume".[3] As far as the *philosophical* essays are concerned (2nd volume) – the editor goes on – the crucial aspect of Meinong's intellectual evolution is the development from a former "psychological, indeed psychologistic"[4] point of view to that of the theory of objects. Now, Höfler claims, the opposite applies to Meinong's *psychological* essays (1st volume): in fact, a comprehensive glance upon them reveals that – "from the very beginning" – these essays "mostly attended theoretical needs".[5] Accordingly, Höfler ideally links Meinong's "psychological" essays to some of the "philosophical" ones: Hume-Studies 1 (1st volume) to Hume-Studies 2 (2nd volume);[6] the essays on complexions and relations and on analysis (1st volume)[7] to that on objects of higher order (2nd volume),[8] and so on.

A discussion of Höfler's criteria is all the more important since they influence the modern standard edition of Meinong's works, started in 1969. Reprinted unaltered, Höfler's two volumes of 1913-1914 make up the first two of the seven volumes (plus supplement) of the *Alexius Meinong Gesamtausgabe*. As the editors

1 Meinong 1913; Meinong 1914. The projected volume *Zur Werttheorie – Vermischtes* (On the theory of value – Miscellaneous) was never published.

2 Höfler 1913, p. v.

3 Höfler 1914, p. ix. Volume 2 appeared in 1913, *before* volume 1.

4 On closer inspection, this phrase favors confusion between two different things. Meinong's abandonment of his early "psychologism" does not necessarily affect his psychology, which he never stopped dealing with. Meinong defines psychologism as "the inappropriate use of psychological method", mostly based "on the neglect or misunderstanding" of the objective side of cognition (Meinong 1904, p. 95–96). See also Meinong 1912.

5 Höfler 1914, p. ix.

6 Meinong 1877, Meinong 1882.

7 Meinong 1891, Meinong 1894.

8 Meinong 1899.

Rudolf Haller and Rudolf Kindinger note in their Preface, added to the Essays on psychology (volume 1), some of the above mentioned "supplements" by Meinong's pupils might be "misleading" from a historical point of view: in fact, they "strove to interpret [...] Meinong's psychological early essays from the point of view of his late developments in the theory of objects, and even to correct them accordingly".[9] As a consequence, future scholarship may run the risk of downgrading these psychological essays as opposed to later ones. Yet – the editors point out – that would be unjustified. The essays included within the first volume have great value of their own, because they exemplify the close kinship between Meinong's "way of philosophizing" and the current "analytic philosophy (of language)".[10] Needless to say, this claim is in line with the broader assumption that the roots of analytic philosophy can be found within the Austrian philosophical tradition, elsewhere developed and defended by Haller.

In sum, we are told that Meinong's essays on psychology are relevant either as preliminary to the theory of objects (Höfler) or as samples of early analytic philosophy of language (Haller). With such claims, both editors end up dismissing a considerable part of Meinong's psychology as irrelevant and – so to speak – almost invisible. Clearly, Höfler's and Haller's editorial choices are not devoid of wisdom, and I am far from suggesting that their interpretative criteria should be disregarded. Editing the writings of a sophisticated thinker like Alexius Meinong is a highly complex task, which involves difficult choices.[11] A strictly chronological edition (preferable, in my view) would have been less biased in suggesting a certain interpretation of Meinong's thought; whereas the available thematic edition is possibly helpful in orienting readers who are not too familiar with the philosopher's work. Anyway, when it comes to assessing Meinong's concept of psychology, the least that can be said is that these editorial criteria call for commentary and – perhaps – a few corrections.

Meinong's compliance with psychologistic stances in his early essays makes his essays on psychology neither irrelevant nor merely preparatory to the theory

9 Haller and Kindinger 1969, p. vii. The supplements have been nevertheless included in the new edition as well, provided that they had been approved by Meinong himself at the time, and that they are neatly separated from Meinong' texts.

10 Haller and Kindinger 1969, p. viii.

11 I fully agree with this statement by Marie-Luise Schubert Kalsi (1996, p. v): "Meinong poses an immense challenge to his interpreters. He did not develop a system. He tackled many problems, he developed intriguing and innovative ideas, and, over the long years of his productive life, he changed his basic philosophical attitude in profound ways. His texts are often obscure, and the interpreter is faced with the task of bringing order into Meinong's thought and with constructing a cohesive system from his ever-changing and diffusive analysis of ideas".

of objects; and the value of those essays does not exclusively reside in the philosophical arguments which forerun the style of analytic philosophy of language. Meinongian psychology is neither a self-deceptive preliminary draft of the theory of objects, nor exclusively an analytic-styled philosophy of language *ante litteram*. To be sure, the "psychological" element of Meinong's production should be understood within the context of the philosophical psychology of the time. As I shall show, however, not all of Meinong's essays in psychology are "philosophical" in equal degree and in the same sense; accordingly, their retrospective interpretation in terms of the theory of objects does not always succeed. Many of Meinong's discussions concern psychological topics which are perfectly in line with the mainstream psychological debates of the time, in a rather broad sense. What is more, such discussions can be found in the essays included by the editors both in the volume on psychology *and* in that on the theory of knowledge and the theory of objects. Höfler's editorial selection is indeed no longer a reliable criterion to single out Meinong's "psychology", both as a thematic field and a methodological stance.

Along with the hitherto discussed editorial problems, two more obstacles stand in the way of a sound understanding of Meinong's psychology: his *style* as a thinker, and his *personal engagement* in psychological research. As to the former question, one should keep in mind that Meinong's style strongly suggests continuity. Reinhardt Grossmann correctly notes that Meinong refrains from sudden changes and – despite his bad reputation of entity multiplier – preferably avoids the introduction of new concepts or terms.[12] Unsurprisingly, in spite of his substantial intellectual evolution, scholars never speak of a "first" and a "second" Meinong. Reluctant to dramatic withdrawals, Meinong rather reformulates and corrects. Accordingly, his early psychology occasionally fades into his theory of objects, especially with the help of the above mentioned "supplements". As I will show, however, this happens in a relatively narrow number of instances. In any case, it would be completely misplaced to argue that Meinong assigned legitimacy to psychology only as a transient phase.

The latter question, concerning his individual engagement, is perhaps thornier. After a certain point in time, Meinong left all the psychological work to his collaborators in Graz. This fact can be interpreted in different ways. Höfler's commentary could mislead one to conclude that Meinong's commitment to the theory of objects eventually made psychology unimportant or superfluous in his

12 Grossmann 1974, p. x.

eyes.[13] By contrast, I argue that Meinong simply applied a labor division strategy. A reformed psychology, consistent with the theory of objects, undoubtedly represents a strategic demand of his mature thought. As known, the turning point is represented by Meinong's distinction between content and object.[14] While opening the fascinating new field of study of objects, the distinction between object and content sheds new light on the previously analyzed characteristics of contents and – at the same time – raises brand new *psychological* problems. Even if there is no late psychological writing by Meinong in the standard edition, then, there is a *Meinongian* psychology before and after the theory of objects, as a part of a scientific program directly inspired by the philosopher.

A supplementary question lies in the relation between theoretical and *experimental* psychology. Meinong must be credited with the foundation of the first experimental laboratory of psychology in the Habsburg Empire, at Graz.[15] Like many others pioneers of the same generation – think of William James – he was of course more inclined to speculation than to the enormously time-consuming experimental work. Speaking as a historian of psychology in the 1950s, Edwin Boring put it quite directly: "Meinong, for all that he founded the first Austrian laboratory at Graz, was a philosopher and not an experimental psychologist".[16] This is basically true, but calls for further explanation. The relation between Meinong's theoretical psychology and the experimental activities in Graz is a question that cannot be evaded with ready-made categorizations. Even though he eventually entrusted pupils like Witasek or Benussi with all experimental activities, Meinong always gave importance to experimentation.[17]

Finally, speaking of Meinong's sources, the most important critical issue is surely that of Brentano's influence.[18] With his lectures in Vienna, Brentano un-

13 See Lindenfeld 1980, p. 220: "[t]o a psychologist, he seemed to have wandered off into the ethereal realms of Gegenstandstheorie and value theory, leaving humble experimenters behind in their laboratories".

14 See below, § 3.

15 See Huber 2012. Meinong himself, already in Vienna, occasionally did some experimental work: Meinong 1921, p. 105.

16 Boring 1957, p. 437. Boring refers to Meinong almost only for his contribution to the debate on "form-qualities".

17 Besides, Meinong also lectured on experimental psychology. According to Lindenfeld (1980, p. 220) "his lectures notes reveal that he expected experiments to be limited to questions of sensations and not to address fundamental theoretical problems"; however, this was not only the case with Meinong, at the time.

18 The influence of Brentano is particularly stressed by Albertazzi / Jacquette / Poli 2001, and Chrudzimski 2007.

doubtedly influenced Meinong no less than many other thinkers of the time. This holds especially for Meinong's earliest writings, which nevertheless bear witness to a marked intellectual independence.[19] In any case, highlighting the importance of Brentano's intellectual heritage should not come at the cost of expunging from the landscape of Meinong's psychology everything that cannot be subsumed under the Brentano-inspired philosophical psychology. The most characteristic concepts of Meinongian psychology – dispositions, activity, and attitudes – have very little, if anything, to do with Brentano's thought and rather utterly oppose its psychological assumptions.

This essay aims at clarifying the role of psychology within Meinong's system of thought, via an assessment of his engagement in it, both in his writings and in the scientific practice in Graz. Bypassing the above mentioned "supplements", I will consider Meinong's early psychology independently of later developments. I will look at the activities of the Graz laboratory as the expression of a consistent scientific plan, directed by Meinong in accordance with theoretical needs emerging form deep within his system of thought. In this way, Meinong's psychological essays and his idea of psychology – before and after the development of the theory of objects – will be proven to be richer and more consistent than what is currently assumed. The following paragraph (§ 2) attempts to illustrate some crucial aspects of Meinong's early psychology in context. I will then proceed to analyze the relations between the theory of objects and psychology in general (§ 3), and to assess the later Meinongian psychology (§ 4).

2 Early Psychological Writings

In his introductory remarks to the 1914 edition,[20] Höfler divides Meinong's psychological writings into the following five groups.

1. I (*Nominalismus*) and VIII (*Abstrahieren und Vergleichen*)
   – about abstraction;[21]

2. V (*Komplexionen*) and VI (*Analyse*)
   – about foundation;[22]

19 See below, § 2.

20 Höfler 1914, p. ix–x. The roman numbers, introduced by Höfler in his edition of 1914, have been maintained unaltered in *Alexius Meinongs Gesamtausgabe*.

21 Meinong 1877, Meinong 1900.

22 Meinong 1891, Meinong 1894.

3. II (*Über Sinnesermüdung*), III (*Empfindung*) and IV (*Phantasie*)
   – about sensory psychology;[23]

4. VII (*Raddrehung*) and IX (*Farbenkörper*)
   – about physiological optics;[24]

5. X (*Urteilsgefühle*)
   – about the basis of ethics.[25]

I will not look into every single essay here. Rather, I will first focus on the first two groups, which are instrumental to the clarification of some critical difficulties and to a comparison with Meinong's later developments. I will then give some hints at the contents of the remaining groups. Two essays taken from the 2nd volume (*Humes-Studien II* and *Über die Bedeutung des Weber'schen Gesetzes*) will be also discussed in this paragraph.[26] Finally, I shall consider some experimental activity carried on by Meinong and his collaborators.

Let us start with the opening piece of the collection, that is, Meinong's habilitation thesis *Hume-Studien I: Zur Geschichte und Kritik des modernen Nominalismus.*[27] Written by Meinong at the age of 24, after meeting Brentano in Vienna, this essay exemplifies an idea of "psychology" that is principally in line with Brentano's stances. Still, Meinong's original thinking and his preference for the analysis of language already emerges.[28] When discussing the existence of abstract ideas, he defends a moderate nominalism.[29] Meinong's essay raises a number of interesting editorial questions, exemplary of the above discussed problems (§ 1). In fact, the second essay *Hume-Studien II: Zur Relationstheorie*[30] was not considered "psychological" enough by Höfler, who included it in the 2nd volume. At first glance, this seems quite odd.

In his commentary to the 1971 edition, Rudolf Haller quotes a number of documents that should support the legitimacy of Höfler's choice. Haller admits that Meinong himself considered his treatment of relations in this essay an example

23 Meinong 1888, Meinong 1888a, Meinong 1889.
24 Meinong 1898, Meinong 1903.
25 Meinong 1905. On this important aspect of Meinong's doctrine, which I will not consider in this essay, see Schubert Kalsi 1996.
26 Meinong 1882, Meinong 1896.
27 Meinong 1877.
28 Haller and Kindinger, p. viii.
29 Barber 1970.
30 Meinong 1882.

of "psychological analysis".[31] Still – he goes on – Meinong's "inadequate [...] psychologization concerns his way of investigating rather than the realm of relations and of their foundations, which Meinong first interpreted as contents of presentation, and later as objects and objects of higher order".[32] This is hardly persuasive: not to mention the circumstance that the same could be said of *Hume-Studien I*, what is at stake is precisely Meinong's "way of investigating". In the wake of Höfler's "supplements", Haller implies that once you replace 'content' with 'object', everything works in this essay. The temptation to read all of Meinong's work from the vantage point of the theory of objects is so pronounced that it induced the editors to scatter over different volumes – and, seemingly, different disciplines – two essays belonging to the same period, bearing the same title (*Hume-Studien*), and whose continuity in method and theme is explicitly pointed out by Meinong himself.[33]

The second group is made up of two essays in theoretical psychology: *Zur Psychologie der Komplexionen und Relationen* and *Beiträge zur Theorie der psychischen Analyse*. Dropping the half-historical approach of *Hume-Studien*, Meinong addresses some of the criticism that many psychologists directed against Christian von Ehrenfels' "gestalt qualities". Interestingly, Meinong introduces a critical remark concerning Ehrenfels' lexical choices:

> Quite by accident I additionally learned that the expression in question is, in some instances, even misleading. I know of one reader who, in connection with this paper, assumed objective, extra-psychic realities behind the "qualities of shape" so that, for example, acoustics had to do not only with sounds, but, moreover, with the melody composed out of these sounds. No wonder, then, that, under these conditions, he had serious objections against the whole theory.[34]

This allegedly accidental remark hints at an important problem of Meinong's later psychology.[35] Meinong concludes: "[w]e are dealing, here, with contents which have [...] a 'basis' and which can be called, in the clearest and most understandable manner, 'founded contents'".[36] By contrast, he will later speak of 'founded objects'[37]

31 Meinong 1882, p. 159.
32 Haller 1971, p. vii.
33 Meinong 1882, p. 5. See also Barber 1971.
34 Meinong 1891, p. 61–62.
35 In the supplements to this essay, Ernst Mally notes: "The 'qualities of shape' are objective, even in the sense of 'extra-psychic', but not real. That is, they cannot exist but must subsist in objects independent of apprehension". (Meinong 1891, p. 71)
36 Meinong 1891, p. 62.
37 See, e. g., Meinong 1899, p. 153, n. 26.

instead. The revision and reformulation in terms of theory of objects, then, succeeds exemplarily in this essay.

Besides, Meinong already highlights here a typical feature of his later psychology: the importance of mental *activity* as a necessary ingredient for the apprehension of complex forms. Ehrenfels believed that mental activity was merely aimed at properly grasping the so-called "basis" of the Gestalt quality: in other words, he thought of it merely in terms of attention. Meinong disagrees:

> It must, of course, be admitted that the founded content accompanies its foundation only under certain conditions. Then, it becomes obvious that in comparing the subject himself adds considerably more than he does in perceiving shape or melody. Even here, not everything is left to the founding contents. In apprehending a melody it is not only important that we hear and reproduce, but also that whatever belongs together remains together, that is, that only certain notes enter into the founding complexion but not other things which precede, follow or even occur at the same time.[38]

Referring to the standard example of melodic perception, Meinong wonders: "There are artistic rules for composition and rendition which should facilitate the listener's task. What would be the purpose of those rules if the listener's task were strictly to listen as attentively as possible?"[39] Furthermore, overcoming Ehrenfels' paradigmatic insistence on the perception of a shape, Meinong stresses the analogy between the perception of *Gestaltqualitäten* and the occurrence of relations in the mind.[40] The difference between the two points of view is striking: whereas perceiving a shape is an effortless process, relating things always implies a certain activity of the mind.[41]

The other essay of this group, *Beiträge zur Theorie der psychischen Analyse*, follows the same line of thought.[42] Written in response to a criticism by Hans Cornelius, Meinong's lengthy and dense essay deserves to stand along with Husserl's *Logical Investigations*, or analogous seminal writings of the time. Meinong defines the core concept as follows: "mental analysis is the limitation of the sphere of judgments",[43] obtained by means of an active increase of what he calls the "weight" of the (partial) representations included in the whole. Once again, then, there is a clear reference to the subject's mental activity.

---

38 Meinong 1891, p. 67-68.
39 Meinong 1891, p. 68.
40 Later he will speak of a "partial coincidence" between them: Meinong 1896, p. 39-40.
41 I shall touch upon this question again: see below, § 4.
42 Meinong 1894.
43 Meinong 1894, p. 104.

In Höfler's subdivision, group 3 is supposed to concern sensory psychology. In fact, the two essays on sensation and on phantasy[44] form a unit of their own. As to phantasy, the core of Meinong's proposal is a rehabilitation of the concept of "disposition", fallen into disgrace because of its misuse within the "psychology of faculties", but potentially useful in a more pragmatic sense. Meinong thinks of psychological concepts like e. g. "sensitivity" (*Empfindlichkeit*), "blunting" (*Abstumpfung*), and "phantasy" (*Phantasie*) in terms of dispositions.[45] Inasmuch as his notion of disposition introduces a modal variation into the subjective "attitude", it lays the groundwork for some later development of his psychology. No wonder, then, that Meinong speaks here of "production" (*Produktion*), and does so in a sense that, despite its generality, foreruns one of the core notions of the mature Meinongian psychology, as Wilhelmine Benussi-Liel points out in her supplement.[46]

The concept of disposition also accounts for Höfler's choice to put these two essays together with *Über Sinnesermüdung im Bereiche des Weberschen Gesetzes* within group 3: "fatigue" (*Ermüdung*) is indeed another dispositional concept for Meinong. In this essay, he matches his analysis of dispositions with a specific method of psychological (or psychophysical) research, which he calls "stimulus reduction" (*Reizreduktion*).[47] Sensory fatigue, Meinong thinks, can be measured in two ways. Suppose that, under normal conditions, a certain stimulus R unleashes a certain sensation E. By definition, when the subject is fatigued, R engenders a sensation e, where $e < E$. Now, psychologists mostly search for the stimulus R' which unleashes E in a fatigued subject (where $R' > R$); by contrast, according to Meinong's method of reduction, they should search for the stimulus r which engenders e in the non-fatigued subject.[48]

It is hard not to have the impression that group 4 is merely a residual container for heterogeneous works. The essay *Über Raddrehung, Rollung und Aberration*[49] is a wonderful example of Meinong's way of linguistic clarification, applied (rather than to philosophy) to the purely *physiological* question of ocular movements. He distinguishes the concepts of aberration, twist and rotation (*Aberration, Raddrehung*, and *Rollung*): the former two respectively express the deviation of the ocular meridian from the eye's absolute meridian and that of the ocular

44 Meinong 1888a, Meinong 1889.
45 Meinong 1889, p. 196.
46 Wilhelmine Benussi-Liel in Meinong 1889, p. 272. See also Stock 1995, p. 459.
47 Meinong 1888, p. 83.
48 Meinong 1888, p. 84. Schematically: in a non-fatigued subject $R \rightarrow E$ and $r \rightarrow e$; in a fatigued subject $R \rightarrow e$ and $R' \rightarrow E$.
49 Meinong 1898.

equator from its equatorial plane; rotation is the deviation from the focus line of the sight.[50] By contrast, despite its title, the essay on *Farbenkörper*[51] is a theoretical work, concerning the abstract representation of colors along a three-dimensional figure, which is not sensory but rather purely intellectual. Meinong admits that the "theoretical (*erkenntnistheoretisch*) point of view" emerges in the essay "more strongly than one could expect, according to the current conventions, in a contribution concerning the theory of colors". He adds:

> After all, I have already practiced enough psychology for the benefit of the theory of knowledge: for the present time, then, I shall be allowed to practice some theory of knowledge for the benefit of psychology, being somehow confident of success.[52]

Subverting Höfler's and Haller's editorial criteria, I will discuss now an important essay in psychology taken from the second volume of Meinong's writings: *Über die Bedeutung des Weber'schen Gesetzes. Beiträge zur Psychologie des Vergleichens und Messens.*[53] Meinong engages in an assessment of the meaning of Weber's psychophysical law – a topic discussed by all the most influential psychologists in the second half of the nineteenth century. To a significant extent, the whole debate on the measurement of mental phenomena pivoted around the distinction between *extensive* magnitudes (e. g. length) and the so-called *intensive* ones (e. g. speed). Needless to say, the issue is of utmost importance for the epistemology of scientific and experimental psychology. In his psychophysics, Fechner treats sensations as intensive magnitudes, and then measurable entities. Many scholars – including Wilhem Wundt – followed this method, which promised to secure a strong epistemic basis to psychology.[54] Meinong's views on the matter are quite original. Leaving technicalities aside, I will sum up his core argument in what follows.[55]

Meinong denies that the opposition between "the intensive" (e. g. a visual sensation) and "the extensive" (e. g. the sensed colored surface) corresponds to the opposition between the mental and the non-mental. Rather, he argues that both intensive and extensive magnitudes occur both outside and inside the mind. This is a remarkable point. That the physical world comprises both categories of magnitudes, extensive and intensive (e. g. length and speed), is

50 Meinong 1898, p. 436.
51 Meinong 1903.
52 Meinong 1903, p. 498.
53 Meinong 1896.
54 I have dealt with this question in Martinelli 1999.
55 For further reference see Tegtmeier 1996, Guigon 2005.

uncontroversial. But the other half of Meinong's claim is astonishing: he is among the few who affirmed that *extensive* magnitudes occur *within* the mind.[56] How did Meinong dare let *res extensa* seep into the *res cogitans*? Did not Descartes distinguish sharply enough between extension on the one side, and the soul on the other? Meinong refers to the two poles of the intentional relation: the act of sensing (*das Empfinden*) is unextended and merely intensive, but "what is sensed" (*das Empfundene*) can be either intensive (e. g. a sound) or extensive (e. g. a length).[57] In his review of Meinong's paper, Bertrand Russell points out:

> To urge, as Herr Meinong does, that imagined space is measurable and divisible, though purely psychical, seems either irrelevant or untrue. For imagined space is as little mental as real space; it differs from real space only in the fact that it does not exist: while the imagination of space, which does exist, is not divisible.[58]

This criticism, later defined by Meinong "completely justified",[59] was a good reason for him to introduce the notion of pseudo-existence, and then to move towards the theory of object.

Together with Stephan Witasek, in 1897 Meinong published an experimental essay on Stumpf's tonal fusion.[60] Since its formulation, Stumpf's *Verschmelzung* has always been central to Meinong's psychology: when the contents (in Meinong's language at the time) are more "blended" (*verschmolzen*), complexions occur instead of relations.[61] In *Zur experimentellen Bestimmung der Tonverschmelzungsgrade* (On the experimental determination of the degrees of tonal fusion) of 1897, stimulated by an experimental research carried on in Graz by his pupil Anton Faist,[62] Meinong resumes a series of experimental attempts made in 1891 about the so-called "degrees" of tonal fusion.[63] Whereas Meinong generally allows for Stumpf's conceptualization, he objects to his colleague's specific determination of the degrees of fusion.[64] The original experimental set was rather

56 Meinong 1896, p. 322. See also Bradley 1895.
57 Meinong 1896, p. 323.
58 Russell 1899, p. 255.
59 Meinong 1899, p. 383, fn. 1.
60 Meinong 1897.
61 Meinong 1891, p. 66. For another discussion see Meinong 1894, p. 112.
62 Faist 1897.
63 See Stumpf 1890, p. 127 ff.
64 Meinong 1891a, p. 167, fn. 2: "Some experiments I have made – just as self-teaching in the beginning – have brought results that diverge from these [*scil.* Stumpf's] statements, at times remarkably".

elementary: Meinong and Witasek, both relatively skilled musicians, alternatively played intervals on the violin, each listening to the intervals while playing. With this, since the violin has no fixed intonation, they aimed at least at the *subjective* belief of the correct intonation of each interval. However, in 1897 they proceed more systematically, using the same apparatus as Stumpf.[65] Commenting on their results, Stumpf welcomed Meinong and Witasek among the few who followed him in investigating tonal fusion empirically, instead of proceeding by mere "conceptual raisonnement".[66] Witasek also performed several experiments with Höfler.[67] A teacher of physics in gymnasia, Höfler had been trained with Ludwig Boltzmann and was an outstanding experimentalist, who cooperated with Meinong to the setting-up of the apparatus that formed the basic equipment of the Graz laboratory.[68]

In sum, the work done in Graz either by Meinong or under his supervision before the turn of the century is basically in line with the standard tasks of a late nineteenth-century laboratory of psychology. In addition to interpreting Weber's law and expressing his view on psychological measurement, Meinong engaged with experimental analyses and theoretic assessments on basic concepts such as sensation, imagination, judgement, abstraction, and analysis, as well as contributions to new problems like that of *Gestaltqualitäten*. Though Meinong's style is markedly theoretical, there is no point in reading *all* of these writings and activities as preliminaries to the theory of objects. This really only applies to a minority of the revised essays: and even in those cases, they prelude to further developments *in psychology* as well.

3 Theory of Object and Psychology

Besides his role as editor of Meinong's writings, Höfler should be mentioned here for at least two more reasons: he was the first to formulate the distinction between object and content and he elaborated the notion of mental "work". Both concepts are crucial for Meinongian psychology.

Though the credit for it often goes to Kazimierz Twardowski's eponymous book of 1894,[69] the distinction between content and object can be found as early

65 Meinong / Witasek 1897, p. 193-194.
66 Stumpf, p. 280.
67 Höfler / Witasek 1900.
68 Höfler 1921, p. 123.
69 See, e. g., Lindenfeld 1980, p. 121.

as 1890 in the *Logik* written by Alois Höfler "with the cooperation" (*unter Mitwirkung*) of Meinong.[70] In any case, Meinong is one of the thinkers who focused the most on this significant distinction, to the point of reformulating his entire scientific program. Admitting that his previous linguistic usage of "object" and "content" was mistaken,[71] Meinong goes so far as to introduce a new discipline, called the theory of objects (*Gegenstandstheorie*). As I have already pointed out, though Meinong focuses on the "objective" (*gegenständlich*) aspect of many of the aforementioned problems, he neither dismisses nor disregards psychology, the discipline in charge for mental contents. Rather, he is left with two *equal-ranking* sets of problems: those pertaining to the object, and those pertaining to the analysis of the content.

When introducing his theory of objects, Meinong stresses its independence on other sciences: metaphysics (ontology)[72] and psychology. He does not deny the proximity between psychology and the theory of objects: indeed, psychology deals (among other things) with objects, as the counterparts of contents. But there are two crucial differences: 1) psychology does not deal with objects *as such*, i.e. *qua* objects; 2) psychology deals (at best) with *some* objects, i.e. only with those that are intended by psychological act (which is not always the case).[73] Let us consider these distinctions more closely. Meinong writes:

> [...] no one fails to recognize that psychological events so very commonly have this distinctive 'character of being directed to something' (*auf etwas Gerichtetsein*) as to suggest very strongly (at least) that we should take it to be a characteristic aspect of the psychological as opposed to the non-psychological.[74]

Clearly, this is the intentionality thesis, introduced by Brentano to distinguish between the mental, or the intentional, and the non-mental or non-intentional. Now, Meinong goes on:

---

70 Höfler / Meinong 1890, p. 6; see Twardowski 1894, p. 4 and Höfler 1921, p. 133. Jacquette 2015, p. 10 speaks of Höfler's and Meinong's "significant abandonment of Brentano's immanence [...] thesis"; by contrast, Chrudzimski 2007, p. 105 believes that Höfler borrows from Brentano. See also Smith 1994, p. 189-190.

71 See, e. g., Meinong 1899, p. 141, referring to content and object: "Even I myself believed for a long time that the expressions could be used interchangeably, that is, that really one or the other of them could be dispensed with. Today I believe this attitude to be unjustified and the linguistic usage, influenced by that attitude, is inexact [...]".

72 I won't discuss here the gap between metaphysics and theory of objects. Suffice it to say that it is parallel to that between Being (*Sein*) and Being-so (*Sosein*).

73 Meinong 1904, p. 88–89.

74 Meinong 1904, p. 77.

> This all-embracing importance of objectivity (*Gegenständlichkeit*) for the psychical side of life – it may indeed be precisely its distinguishing characteristics – as I have briefly mentioned already, may now suggest to us that we [...] permitted ourselves to be led down a detour that might easily been avoided. For the science which most naturally would have to do with objects as such would be the very one whose business is to deal with this objectivity. This task [...] seems to belong to psychology.[75]

If intentionality is the "distinguishing characteristic" of the mental, and intentionality is understood as directness to objectivity, it seems very hard to escape the conclusion that mental life and objectivity are closely related to each other: accordingly, psychology may seem to be in charge for objects and objectivity as well. Yet this approach would fail to cover a number of problems. As previously said, psychology deals with objects only indirectly, that is, it does not deal with objects *as such* (see above, 1). Only the theory of objects considers them for their own sake, and not in guise of psychological explanations. Furthermore, with the theory of objects Meinong also considers those objects that do *not* enter in any intentional relation (see above, 2), eschewing the domain of psychology altogether.[76]

Of the two main differences singled out by Meinong, only the latter points at a dimension where theory of objects and psychology are entirely independent of each other. By contrast, the former point ("psychology does consider objects, but not as such") shows that psychology is not really dispensable for Meinong. The most interesting problems, of course, are those concerning *ideal* objects and the relationship between intentionality and non-existing objects.

> According to modern usage [...] the word "ideal" means the same as "thought of" or "merely presented"; hence it pertains, apparently, to all of those objects which do not exist or which could not exist. What does not exist outside of us, so one automatically thinks, must at least exist in us. Such an object, it is supposed, belongs before the forum of psychology; one then makes room for the thought that the knowledge of existing things (and along with this knowledge reality itself) can perhaps be treated "psychologically".[77]

Carving out a role for the newly introduced discipline, Meinong comprehensibly insists on the limitations of psychology and stresses the independence of the theory of objects on the latter. Yet psychology clearly retains a definite and important role.

---

75 Meinong 1904, p. 87.
76 That there are objects of this kind – and how are we supposed to know about them – is explained by Meinong in detail; but following him would take us too far.
77 Meinong 1904, p. 96.

Scholars acquainted with Meinong's theory of objects are familiar with his peculiar terminology. Meinong speaks of "founded objects" or "higher order objects". Accordingly, a higher order object (*Superius*) is "founded" on the elements that compose it (*Inferiora*). Thus, the "foundation" turns out to be a crucial process both for the theory of objects and for epistemology in general: even the core principle of empiricism must be integrated with the remark that the foundation discloses the entire world of ideal objects.

> The old epistemological-psychological principle of intellectus and sensus suffers from the fundamental defect in that it neglects, in the face of foundation, the other half of facts.[78]

Now, there is a parallel process on the psychological side, but Meinong insists on the differences between the two instances. As a result of the acquisition discussed above, he concludes that there is a founded object, but no founded *content*. Founded objects, rather, have their counterparts in what Meinong and his followers call "produced" contents.[79] In other terms, *production* (*Produktion*) is the mental process that corresponds to foundation: *founded* objects have their psychological counterpart in *produced* contents (*produzierte Inhalte*). The investigation of mental "production" leaves Meinong with a wide field open to intriguing psychological investigations, which has been only partially exploited by scholars so far.[80] In the following paragraph, I will draw attention to the evolution of Meinongian psychology after the acquisition of the distinction between object and content and the development of the theory of objects.

4 Late Meinongian Psychology

Some doubts concerning Meinong's construct might arise at this point. Is *Produktion* a mere reduplication, a "projection" of foundation onto psychological contents? After all, Meinong's fame as entity multiplier may speak for this hypothesis. However, Meinong considers two main differences between the processes of foundation and production: 1) psychological contents are not ideal, but rather real; 2) in the case of production there is a true psychological *activity*. The former

78 Meinong 1899, p. 153.

79 Ameseder 1904, p. 486–488.

80 Albertazzi 2001, p. 239 notes that so far there has been "no detailed reconstruction of the theory [*scil.* of production]". In my view, this follows from the vicarial and unclear status ascribed to psychology within the Meinongian scholarship.

claim points at the ontological nature of psychological contents. Psychology always deals with realities, that is, with mental acts and contents. Foundation, by which ideal objects emerge, has no place within psychology as such. With this, the distinction between objects and contents displays its theoretical potentialities. From the point of view of those who welcome the natural bias "in favor of real objects" (rather than considering it a "prejudice", as Meinong does)[81] one could even say that the new conceptual arrangement imposes to *confine* ideality within the theory of objects, thus freeing psychology from it. This goes hand in hand with the latter claim, concerning mental activity. The mental process of production, in fact, is not automatic: it absorbs a quantum of energy. These two pedestals lay the epistemological foundations of the mature Meinongian psychology.[82]

In 1904, ten years after the foundation of the Graz laboratory, Meinong edited a celebrative volume with the title *Untersuchungen zur Gegenstandstheorie und Psychologie*. The volume includes eleven contributions authored by Meinong, Rudolf Ameseder, Ernst Mally, Wilhelm Frankl, Vittorio Benussi, Wilhelmine Liel and Robert Saxinger. The first three essays (by Meinong, Ameseder and Mally) deal with the theory of objects, whereas the remaining ones tackle psychological themes. The volume has extraordinary importance as an example of the collective working program around Meinong.[83]

A synthesis of the interests and methods of this school is given by Wilbur Marshall Urban in his ample review of the volume.

> Psychology is the science not of any objects as such but of the processes involved in the experience of those objects, sensation, perception, presentation, judgment, assumption, feeling, desire, etc., directed upon these objects, through which the objects are presented, grasped, constructed, valued. These studies are not based upon that conception of psychology which finds its chief function in the analysis of objects, i.e. in the exploration of the manifold of sensation objects, colors, tones, etc. [...], but rather upon that which finds its chief interest in the processes by which objects of higher order are constructed on the basis of the simples of sensation and perception. Psychology has to do with one's attitude (*Verhalten*), not with objects. And with this comes the use of [...] "psychical activity" or energy, that some commentator would rather banish from psychology. And these concepts of attitude and psychical activity are actually used as

81 Meinong, 1904, p. 78.

82 As previously noted, it would not be completely appropriated to speak of *Meinong's* later psychology, because Meinong eventually concentrated his efforts onto the theory of objects, leaving psychological undertakings to his collaborators.

83 Meinong promoted a cooperative view in scientific and philosophical work, which made the Graz group a "school" only in the best sense of the term, with no dogmatic implications.

> explanatory principles in accounting for certain facts of analytical and experimental analysis.[84]

Urban draws attention to the most important features of the reviewed psychology: the presence of mental activity and the role assigned to mental attitudes.[85]

Let us begin with mental activity. The foundations for this theory had been laid by Höfler in an essay devoted to "mental work" (*psychische Arbeit*). In analogy with physics, Höfler distinguishes work from energy, the former concept being defined as the product of force and distance.[86] Interestingly, Höfler does not indulge in a special (and complicated) mental mechanics *à la* Herbart, but rather considers whether the very *same* laws of physical work apply within the realm of mental activity. The notion of mental work is applied onto the fields of will, emotions and feelings, judgement, presentation, etc. For instance, Höfler insists on the concept of "mass" as a suitable substitute for Meinong's "weight" of presentations.[87] In sum, Höfler's essay suggests that no mental activity goes without a certain amount of mental work, which is potentially measurable.

This conceptual framework sits well with the Meinongians' insistence on "attitudes". Meinong and his pupils noted that subjects can either produce higher-order contents or not, according to the general circumstances and to the specific attitude of the mind. For instance, in his analysis of the Müller-Lyer figure, Benussi individuates two kinds of "reaction" that lead the subject respectively to consider the perceptual stimulus as a whole, or to analyse it. Benussi calls these two attitudes 'G-Reaktion' and 'A-Reaktion', where G means *Gestalt* and A means *analysis*.[88] Benussi investigates the insurgence of the two reactions experimentally. He concludes that the deceptive character of the Müller-Lyer figure is the result of an illusion due to the process of "production" of the complex representation (*Produktionstäuschung*).[89] The deception originates neither in sensation nor in judgment: the problem affects the process of production.

For all its importance, the 1904 volume is far from exhausting the richness of Meinongian psychology. Here one ought to mention at least one other outstanding figure, already referred to above: Stephan Witasek. In an essay on *Komplex-*

---

84 Urban 1906, p. 69-70.

85 *Verhalten* could be perhaps better translated as 'behavior', but that would be misleading in suggesting some sort of kinship with behaviorist psychology.

86 Höfler 1895, p. 46.

87 Höfler 1895, p. 162 ff. On Meinong's concept of "weight" see and Meinong 1894 and above.

88 Benussi 1904, p. 309.

89 Benussi 1904, p. 412.

ionen und Relationen which integrates Meinong's study of the topic, Witasek insists on the presence of different attitudes as well. For instance, the proper manner of listening to polyphonic music implies a complete distinction between all voices from one another, but there are various degrees of approximation to that ideal case, depending on the quality of the execution and on the musical education of the listener.[90]

Witasek was in charge of the Graz laboratory at the time of his untimely death in 1915, at age 45. The difficulties of establishing a stable direction for the laboratory was an obstacle for the development of experimental activities in Graz.[91] Benussi was the natural candidate for Witasek's role, but the outburst of World War I made his situation – as an Italian in Austria – quite delicate. He eventually left and went back to his homeland, where he founded one of the most important traditions of experimental psychology in Padua.[92] These difficulties massively contributed to the decline of the Graz laboratory.

In conclusion, Meinong and the other psychologists of the Graz school share a robust epistemic view and an ambitious program of experimental activity, consistent with Meinong's theoretical speculation, partly inspired by it and fruitfully in line with it. The distinction between objects and contents, which are in turn related to mental acts, has capital importance here. Psychology is principally interested in acts and contents. Contents do not obey the same laws as objects: foundation is different from production, which is a real process that absorbs *quanta* of energy. The manifold "dispositions" and the different subjective "attitudes" which lead to a regular, lawful deviation from standard response, can and should be investigated experimentally. Despite being an independent discipline, psychology nevertheless fits into a full-fledged system of philosophical assumptions. Though psychologism had no part in it (at least in the final version of the doctrine), Meinong and his pupils shared the view that psychology was important for philosophical disciplines. As Meinong had already noted in 1885,

> Psychology [...] is not philosophy altogether; rather, it is a part of it, its fundamental discipline: in fact, the objects that psychology elaborates make up the material that give every philosophical consideration its peculiar characteristics.[93]

---

90 Witasek 1897, p. 420.

91 Huber 2012, p. 216, 225.

92 See Antonelli 1994, 2018.

93 Meinong 1885, p. 5. See Marek 2013: "Although Meinong had constantly recognized the fundamental role of psychology in philosophy" he increasingly became an "opponent of psychologism".

References

GA = Meinong, Alexius, *Alexius Meinong Gesamtausgabe*, 7 vols. and supplementary volume, ed. by Rudolf Haller et al., Graz: Akademische Druck- und Verlagsanstalt 1969–1978.
UGP = *Untersuchungen zur Gegenstandstheorie und Psychologie. Zum zehnjährigen Bestande des Psychologischen Laboratoriums der Universität Graz*, hrsg. von A. Meinong, Leipzig: Barth, 1904.
OHO = Meinong, Alexius, *On Objects of Higher Order and Husserl's Phenomenology*, ed. by Marie-Luise Schubert Kalsi, The Hague: Nijhoff, 1978.

Albertazzi, Liliana (2001), "Presentation and Production", in: *The School of Alexius Meinong*, ed. by L. Albertazzi, D. Jacquette and R. Poli, London: Routledge, p. 239–260.
Albertazzi, Liliana / Jacquette, Dale / Poli, Roberto (2001), *The School of Alexius Meinong*, London: Routledge.
Ameseder, Rudolf (1904), "Über Vorstellungsproduktion", in: UGP, p. 481–508.
Antonelli, Mauro (1994), *Vittorio Benussi. Leben und Werk*, Amsterdam: Rodopi.
Antonelli, Mauro (2018), *Vittorio Benussi in the History of Psychology. New Ideas of a Century Ago*, Dordrecht: Springer.
Barber, Kenneth (1970), "Meinong's Hume Studies: Part I: Meinong's Nominalism", in: *Philosophy and Phenomenological Research* 30/4, p. 550-567.
Barber, Kenneth (1971), "Meinong's Hume Studies: Part II. Meinong's Analysis of Relations", in: *Philosophy and Phenomenological Research* 31/4, p. 564-584.
Benussi, Vittorio (1904), "Zur Psychologie des Gestalterfassens (Die Müller-Lyersche Figur)", in: UGP, p. 303–448.
Boring, Edwin (1957), *A history of experimental psychology* (2nd ed.), New York: Appleton-Century-Crofts.
Bradley, Francis Herbert (1895), "In What Sense Are Psychical States Extended?", in: *Mind* 14 (n.s.) p. 225.
Brentano, Franz (1907), *Untersuchungen zur Sinnespsychologie*, ed. by R. Chisholm e R. Fabian, Hamburg: Meiner, 1979.
Chrudzimski, Arkadius (2007), *Gegenstandstheorie und Theorie der Intentionalität bei Alexius Meinong*, Dordrecht: Springer.
Faist, Anton (1897), "Versuche über Tonverschmelzung", in: *Zeitschrift für Psychologie und Physiologie der Sinnesorgane* 15, p. 102-131.
Grossmann, Reinhardt (1974), *Meinong*, London: Routledge & Kegan Paul.
Guigon, Ghislain (2005), "Meinong on magnitudes and measurement", in: *Meinongian Issues in Contemporary Italian Philosophy*, ed. by A. Schramm, Frankfurt a.M.: Ontos, p. 255–296.
Haller, Rudolf (1971), "Vorwort", in: GA, vol. 2, p. vii–xi.
Haller, Rudolf / Kindinger, Rudolf (1969), "Vorwort", in: GA, vol. 1, p. vii–xi.
Höfler, Alois (1895), "Psychische Arbeit", in: *Zeitschrift für Psychologie und Physiologie der Sinnesorgane* 8: p. 44–103, 161–230.
Höfler, Alois (1913), "Vorwort zum ersten und zweiten Band", in: *A. Meinong's Gesammelte Abhandlungen. Herausgegeben und mit Zusätzen versehen von seinen Schülern*, Leipzig: Barth, p. v–vii; repr. in GA, vol. 1, p. 619–621.

Höfler, Alois (1914), “Zur Orientierung über den Inhalt des ersten Bandes”, in: *A. Meinong's Gesammelte Abhandlungen. Herausgegeben und mit Zusätzen versehen von seinen Schülern*, Leipzig: Barth, 1914, p. ix–x.

Höfler, Alois (1921), “Die Philosophie des Alois Höfler [Selbstdarstellung]”. In: Schmidt, R., Hrsg., *Die Deutsche Philosophie der Gegenwart in Selbstdarstellungen*, Leipzig: Meiner, p. 117–160.

Höfler, Alois / Meinong, Alexius (1890), *Logik*, Wien-Prag: Tempsky, Leipzig: Freytag.

Höfler Alois / Witasek, Stephan (1900), *Psychologische Schulversuche mit Angabe der Apparate* Leipzig: Barth [1903^{2} and following editions: *Hundert Psychologische Schulversuche mit Angabe der Apparate*].

Huber, Helmuth P. (2012), „Das Grazer „Psychologische Laboratorium“ um 1900. Methoden, Konzepte, Forschungsthemen“, in: *Psychologische Rundschau*, 63 (4), p. 218–227.

Jacquette, Dale (2015), *Alexius Meinong, the Shepherd of Non-Being*, Dordrecht: Springer.

Kries, Johannes von (1882), “Über die Messung intensiver Grössen und über das sogenannte psychophysische Gesetz”, in: *Vierteljahrsschrift für wissenschaftliche Philosophie* 6, p. 257–294.

Lindenfeld, David (1980), *The Transformation of Positivism: Alexius Meinong and European Thought*, Berkeley: University of California Press.

Marek, Johann (2013), “Alexius Meinong”, in: *The Stanford Encyclopedia of Philosophy* (Fall 2013 Edition), Edward N. Zalta (ed.), URL = <https://plato.stanford.edu/archives/fall2013/entries/meinong/>.

Martinelli, Riccardo (1999), *Misurare l'anima. Filosofia e psicofisica da Kant a Carnap*, Macerata, Quodlibet.

Meinong, Alexius (1877), „Hume-Studien I: Zur Geschichte und Kritik des modernen Nominalismus“, in: GA, vol. 1, p. 1–108.

Meinong, Alexius (1882), “Hume-Studien II: Zur Relationstheorie”, in: GA, vol. 2, p. 1–76.

Meinong, Alexius (1885), “Über philosophische Wissenschaft und ihre Propädeutik”, in: GA, vol. 5, p. 1–196.

Meinong, Alexius (1888), “Über Sinnesermüdung im Bereiche des Weberschen Gesetzes”, in: GA, vol. 1, p. 79–108.

Meinong, Alexius (1888a), “Über Begriff und Eigenschaften der Empfindung”, in: GA vol. 1, p. 109–192.

Meinong, Alexius (1889), “Phantasie-Vorstellung und Phantasie”, in: GA vol. 1, p. 193–277.

Meinong, Alexius (1891), “Zur Psychologie der Komplexionen und Relationen”, in: GA, vol. 1, p. 279–303. Quoted from the English translation “On the Psychology of Complexions and Relations”, in: OHO, p. 55-71.

Meinong, Alexius (1891a), “Rezension von: Carl Stumpf, *Tonpsychologie*, Zweiter Band. Leipzig, Verlag von S. Hirzel, 1890. 8 XIV und 582 Seiten”, in: GA, vol. 7, p. 163-174.

Meinong, Alexius (1894), “Beiträge zur Theorie der psychischen Analyse”, in: GA, vol. 1, p. 305–395. Quoted from the English translation “An Essay Concerning the Theory of Psychic Analysis”, in: OHO, p. 73-135.

Meinong, Alexius (1896), “Über die Bedeutung des Weber'schen Gesetzes. Beiträge zur Psychologie des Vergleichens und Messens”, in: GA, vol. 2, p. 215–376.

Meinong, Alexius (1898), “Über Raddrehung, Rollung und Aberration”, in: GA, vol. 1, p. 397–441.

Meinong, Alexius (1899), "Über Gegenstände höherer Ordnung und deren Verhältnis zur inneren Wahrnehmung", in: GA, vol. 2, p. 377–471. Quoted from the English translation "On Objects of Higher Order and their Relation to Internal Perception", in: OHO, p. 73-135.

Meinong, Alexius (1900), "Abstrahieren und Vergleichen", in: GA, vol. 1, p. 443–494.

Meinong, Alexius (1903), "Bemerkungen über den Farbenkörper und das Mischungsgesetz", in: GA. vol. 1, p. 495–576.

Meinong, Alexius (1904), "Über Gegenstandstheorie", in UGP, p. 1–50; repr. in: GA, vol. 2, p. 481–537. Quoted from the English translation: "The Theory of Obiects", transl. by Isaac Levi, D.B. Terrella and Roderick M. Chisholm, in: *Realism and the Background of Phenomenology*, ed. by R.M. Chisholm, Chicago: Free Press of Glancoe, 1960.

Meinong, Alexius (1905), "Über Urteilsgefühle, was sie sind und was sie nicht sind", in: GA, vol. 1, p. 577–615.

Meinong, Alexius (1912) "Für die Psychologie und gegen den Psychologismus in der allgemeinen Werttheorie", in: GA, vol. 3, p. 269–282.

Meinong, Alexius (1913), *A. Meinong's Gesammelte Abhandlungen. Herausgegeben und mit Zusätzen versehen von seinen Schülern*. Bd. 2: *Abhandlungen zur Erkenntnistheorie und Gegenstandstheorie*, Leipzig: Barth.

Meinong, Alexius (1914), *A. Meinong's Gesammelte Abhandlungen. Herausgegeben und mit Zusätzen versehen von seinen Schülern*. Bd. 1: *Abhandlungen zur Psychologie*, Leipzig: Barth.

Meinong, Alexius (1921), "Alexius Meinong", in: GA, vol. 7, p. 1–62.

Meinong, Alexius / Witasek, Stephan (1897), "Zur experimentellen Bestimmungen der Tonverschmelzungsgrade", in: *Zeitschrift für Psychologie* 15, p. 189–205.

Meister, Richard (1959), "Ehrenfels, Maria Christian Julius Leopold Karl Freiherr von", in: *Neue Deutsche Biographie*, vol. 4, p. 352 f. [online version: https://www.deutsche-biographie.de/gnd118529277. html#ndbcontent].

Russell, Bertrand (1899), "Review of Meinong, Über die Bedeutung des Weber'schen Gesetzes", in: *Mind* 8 (n.s.), p. 251–256; also in: *The Collected Papers of B. Russell*, vol. II: *Philosophical Papers* (1896-99), ed. by N. Griffin e A.C. Lewis, London, 1990, p. 148–152.

Schubert Kalsi, Marie-Luise (1996), *Alexius Meinong's Elements of Ethics. With Translation of the Fragment Ethische Bausteine*, Dordrecht, Kluwer.

Smith, Barry (1996), *Austrian Philosophy. The Legacy of Franz Brentano*, Chicago: Open Court.

Stock, Wolfgang G. (1995), "Die Genese der Theorie der Vorstellungsproduktion der Grazer Schule", in: *Grazer Philosophische Studien* 50, p. 457-490.

Stumpf, Carl (1890), *Tonpsychologie*, vol. 2, Leipzig: Barth.

Stumpf, Carl (1898), "Neueres über Tonverschmelzung", in: *Zeitschrift für Psychologie* 15, p. 280–303.

Tegtmeier, Erwin (1996), "Meinong on Measurement", in: *Grazer philosophische Studien* 52, 1996/97, p. 161-171.

Twardowski, Kazimierz (1894), *Zur Lehre vom Inhalt und Gegenstand der Vorstellungen. Eine psychologische Untersuchung*, Wien: Hölder.

Urban, Wilbur Marshall (1906), "Review of: Untersuchungen zur Gegenstandstheorie und Psychologie by A. Meinong", in: *The Philosophical Review* 15/1, p. 65–75.

Witasek, Stephan (1897), "Beiträge zur Theorie der Komplexionen und Relationen", in: *Zeitschrift für Psychologie* 14, p. 410–435.

Denis Seron

Color Geometry – Or Color Grammar?

Abstract: This article discusses some difficulties of the theory of color propounded by Meinong in his *Remarks on the Color Solid and the Mixture Law* of 1903. First, I argue that Meinong's geometrical approach faces at least three sets of difficulties related to the following assumptions: colors possess a "nature" that can be grasped through intuition; they are separated from each other by continua in color space; there are an infinite number of a priori relations between colors. Second, I confront the geometrical approach with Wittgenstein's grammatical approach, contending that the latter escapes these difficulties.

Introduction

The present paper is about the epistemology of the theory of color and aims to confront two different approaches. In the first half, I present an overview of some key tenets of Meinong's *Remarks on the Color Solid and the Mixture Law* (*Bemerkungen über den Farbenkörper und das Mischungsgesetz*) of 1903. In the second half, I discuss Meinong's approach at length and compare it with another approach which I tend to favor, namely the grammatical and pragmatic approach promoted by Wittgenstein. I highlight some of its difficulties and raise some objections.

1 Meinong's Color Geometry

The question I will address is whether the theory of color can be categorized as an a priori science, and, if so, in what sense. At first glance, Meinong's position in this respect may seem somewhat schizophrenic. On the one hand, he insists that the psychological approach to colors is highly misleading. As he claims on the opening pages of his 1903 paper, "colors are as little mental as are places or even numbers".[1] He appeals for an a priori theory of color that is completely autonomous from

1 Meinong 1903, p. 3.

Denis Seron, Département de Philosophie, Université de Liège
E-mail: d.seron@uliege.be

https://doi.org/10.1515/9783110664850-003

empirical psychology – a theory of color that is more like mathematics and his theory of objects. His claim is that the theory of color must be analogous to Euclidean geometry. On the other hand, however, Meinong maintains that empirical findings must play a central role in the theory of color and that, in some sense, even the geometrical representation of the relations between colors belongs to empirical psychology.

In fact, Meinong's purpose is precisely to clarify the relationship between what is empirical and what is a priori in the theory of colors – and this is what makes his *Remarks on the Color Solid* philosophically interesting and illuminating.

Meinong's most central claim about the theory of color is twofold. First, the theory of color is, at least in part, an a priori science, namely that science which studies a priori relations among colors, for example the relations "orange is between red and yellow" or "no color is composed of two other colors". That the theory of color is an a priori science means, for Meinong, that it is not about the existence of particular colors or colored things, but about the "nature" (*Natur*, *Beschaffenheit*) of colors, namely about internal relations among colors.[2] As such, the theory of color has to do with his theory of objects and, more specifically, with his theory of higher-order objects.[3]

The second claim is that these internal relations among colors can be schematized using a three-dimensional solid. In other words: each color is completely determined or defined by a triplet of coordinates just as are points in Euclidean geometry.

Let us draw three perpendicular lines. The end points of the first line – say, the x-axis – are pure red and pure green. The end points of the second line – the y-axis – are pure blue and pure yellow. Those of the third line – the z-axis – are white and black. All visible colors, that is, all the colors that are inside the color solid, must be representable as a triplet of coordinates on the three axes. For example: pure green and pure red have only one value in common, namely that on the white-black axis; purple has one value in common with some hue of blue and one value in common with some hue of red, and so on. Since every color is defined by its coordinates, it is a priori impossible for two different colors to be at the same place in the color space. Two different colors that would have all of their coordinates in common would be both different and identical. In Meinong's view, this means that the compositional view of color is false. For example, purple is not a compound of red and blue. However, if we take the word "compound" in a

2 Meinong 1903, p. 3; Mulligan 1991, p. 80.
3 Meinong 1903, p. 4.

looser sense, then we must say that all colors are compounds insofar as they are analyzable in three coordinates.[4]

To sum up: the theory of color is an a priori science and it deals with relations that are analogous to geometrical relations. The immediate consequence of these two claims is that the theory of color should be conceived of as a "color geometry" (*Farbengeometrie*).

The idea that color relations are analogous to geometrical relations and hence representable through geometrical figures was not new at the time Meinong wrote his *Remarks on the Color Solid*. It had been at the basis of color wheels such as those proposed by Goethe and some others already in the eighteenth century. Most color theorists in the nineteenth and twentieth centuries opted, like Meinong, for three-dimensional models. Among them are Helmholtz's color cone, Ostwald's double cone, the color sphere of Runge and Wundt, the octahedron of Höfler, Ebbinghaus, and Wittgenstein.[5] In Höfler's color octahedron, for example, the existence of a line from yellow to red indicates the possibility of orange hues; the absence of any line between red and green indicates the impossibility of reddish green; and so on.

For some reasons Meinong seems to prefer Ebbinghaus's color octahedron,[6] but this is of secondary importance. What is important here is the epistemological analogy between geometry and color theory, namely the view that both provide a priori knowledge. More precisely, the question Meinong is primarily concerned with – and the question I will focus on at a first stage – is about the relationship between color theory and empirical psychology. Put quite simply, the question is how we are to reconcile the view that the theory of color provides a priori knowledge with the fact that psychological experiments are obviously relevant to the theory of color.

The color solid has many properties that are known through a priori intuitions (*apriorisch einzusehen*) and can be expressed as a priori laws. For example, it is immediately self-evident that red is more distant from green than from orange, that you do not need to change direction in order to go from red to gray and hence that the line between them must be a straight line, etc.[7] However, it is plausible to say that not all relations represented in the color solid are a priori relations. Meinong considers the bounds of the solid. For example: the most distant colors from neutral gray on the red-green line are pure red and pure green, so the

4 Seron 2015; Seron 2016.

5 For an overview of those debates, see Kuehni / Schwarz 2008, Crone 1999.

6 Meinong 1903, p. 8.

7 Meinong 1903, p. 6.

red-green line is geometrically represented as a line segment whose two end points in the color solid are pure red and pure green. In Meinong's view, the fact that pure red, green, blue and yellow are end points is the only sense in which they can be characterized as "basic" colors.[8]

The question is, Is the statement that the red-green line is bounded by pure red and pure green an a priori truth? Or to put it otherwise: Is it a priori self-evident that the red-green line is bounded by pure red and pure green? Meinong's answer is no. The colors that are beyond the bounds of the octahedron are not a priori impossible. These colors – which I will from now on call *transcendent colors* (the term is mine) – are possible even though *we* cannot perceive them. On the one hand, our inability to perceive transcendent colors is a contingent fact due to factors such as the structure of our sense organs, which are subject to empirical inquiry. On the other hand, we do have a priori intuitions about colors that are inaccessible to us.[9] For example, it is a priori self-evident that gray is between transcendent red and transcendent green, or that all transcendent colors on the green-red line are such that you do not need to change direction in order to go from one to another.

The question is not without a certain poetry. It is whether it makes sense to talk about colors that are outside the color solid and thus inaccessible to us, for example a blue which is bluer than pure blue, a green which is greener than pure green. Meinong's answer is that it does: it is a priori possible for there to be transcendent colors and these transcendent colors, like the visible ones, fall within the domain of a priori color theory.

By contrast, the bounds and thus the shape of the color solid are empirical. The fact that the maximal values on the red-green line are pure red and pure green is an empirical fact, open to empirical investigation.

The upshot of this is that the theory of color must have an a priori part and an empirical part. The a priori part is that branch of the theory of objects which deals with what colors are – with their "nature" as it can be known through a priori intuitions. The empirical part belongs to psychology and is about colored things as contents of sensory presentations.[10] Transcendent colors are studied in the a priori part of color theory, not in its empirical part.

Meinong captures this distinction by saying that the empirical or psychological part of color theory deals with actual existence, while its a priori part deals with a priori possibility. This, of course, applies to the color solid's bounds. It is

8 Meinong 1903, p. 34.
9 Meinong 1903, p. 8.
10 Meinong 1903, p. 4.

an empirical fact that your psychological constitution enables you to perceive such-and-such colors and not other possible colors. "The end point, Meinong argues, is a psychological datum. The fundamental delimitation <of the color solid> is about something actual, not something possible".[11] As he argues some pages earlier, "the fact that the intuitions which are at the basis of the color solid and made manifest by the color solid's shape count as a piece of psychology is doubtless to be explained by the relatively limited operational abilities of our intellect in the field of colors".[12]

To sum up: the color solid itself must be somehow grounded in empirical facts. As Meinong claims, "the color solid cannot be viewed as a result of mere a priori knowledge; (...) it must in part be legitimated by empirical evidence (*durch die Empirie legitimiert*)".[13] In this sense, the color solid deserves to be called a "*psychological* color solid".[14]

Given this, the questions now are: Why is the color solid not only psychological? What is properly a priori in the color solid? What is the difference between what is empirical and what is a priori in the color solid? Most illuminatingly, Meinong appeals to the analogy with geometrical space. The idea is that the color solid is part of a color space just as, say, your hand is part of the Euclidean space. Your hand exhibits many geometrical properties and relations that are accessible to a priori intuition and subject to geometrical laws. However, its shape is obviously something empirical. Likewise, the color solid, as part of the color space, displays properties and relations that are subject to a priori intuition and knowledge. Thus, the difference between the empirical and the a priori parts of the theory of color coincides with the difference between the color solid and the three-dimensional color space it is part of.[15] As Meinong summarizes:

> Like every real solid, the color solid, too, is in space and participates in the properties of space; but if I am right, it could help overcome some difficulties to very explicitly characterize the space here in question as a *color space*, and to oppose it, as the proper object of a priori color knowledge, to the color solid taken as the object of the corresponding observations, which are in principle empirical.[16]

11 Meinong 1903, p. 9.
12 Meinong 1903, p. 3.
13 Meinong 1903, p. 10.
14 Meinong 1903, p. 11.
15 Rollinger 2008, p. 125.
16 Meinong 1903, p. 12.

The emerging picture can be summarized as follows. The color solid presents all the colors that are accessible to human experience. The corresponding empirical data are materials for a priori intuitions about the nature – or internal relations – of visible colors.[17] However, as Meinong emphasizes, "the a priori treatment, here as anywhere else, extends beyond that which experience presents to it".[18] The color space can be viewed as a geometrical representation of such an "a priori treatment".[19]

2 Color Geometry vs. Color Grammar

Let us now attempt a brief comparison between Meinong's color geometry and a grammatical approach in the style of Wittgenstein. I will not delve into the details and will confine myself to a couple of sketchy remarks on some differences that I consider most significant from a philosophical point of view.

Consider the following view, leaving aside the question whether it is the actual view of Wittgenstein or any other: (1) the theory of color includes a priori laws such as "orange is between red and yellow"; (2) all of these laws are purely conceptual in some specific sense: they are rules for the use of color words. The a priori part of color theory is about our practice with color words.

To some extent, this view is not so very different from Meinong's view. It is well known that Wittgenstein himself accepted the idea of a geometrical representation of the internal relations among colors.[20] In his *Philosophical Remarks*, he hesitates between Höfler's octahedron and Ostwald's double cone.[21] The similarity is even more striking if Meinong's idea of analyzing colors in three coordinates is understood, say, in terms of conceptual analysis. In the *Tractatus*, 6.3751, and later in his *Some Remarks on Logical Form*, Wittgenstein endorses the view that each color can be logically analyzed in a sequence of numbers.[22] In the *Philosophical Remarks*, he clearly assumes that color solids are *grammatical* representations.[23]

17 Meinong 1903, p. 4.
18 Meinong 1903, p. 4–5.
19 Cf. Rollinger 2008, p. 127: "When Meinong speaks of a priori knowledge he means only that it is to be obtained strictly from a consideration of the nature of the known object. This obviously does not rule out the possibility that the concept of what is known has its origin in experience".
20 Mulligan 1991, p. 82; Carvalho 2017, p. 321 ff.
21 Wittgenstein 1975, p. 278; Barceló / Saab 2017.
22 Wittgenstein 1963, p. 144–5; Wittgenstein 1929, p. 166–7.
23 Wittgenstein 1975, p. 51; Park 1998, p. 136–42; cf. Wittgenstein 1977, p. 23.

Nevertheless, there are significant differences between the two views. I mention three of them that strike me as more salient:

(1) The first one concerns Meinong's essentialism, namely his idea that the coordinate triplets in the color solid capture the "nature" of colors and that this nature is accessible to a priori intuition. At the most general level, there is an enormous difference between understanding what the word "orange" means and having an a priori intuition of the "nature" of orange color. In order to understand what a word means you need, say, to know in what circumstances it ought to be used. By contrast, the notion of intuition suggests that a priori knowledge is more like seeing something that resides in individual things.

(2) The second difference is that, according to Meinong's analogy with geometry, the color space has a character of continuity which grammatical structures do not have. In the color octahedron, pure red and pure yellow are connected by something analogous to a geometrical line, which indicates that there is a continuum between them. Of course, you can express the same idea by saying that the distances between colors are continuous magnitudes, or that each color is defined by three variables with any real numbers as values. By contrast, the distances between color words or concepts are neither continua nor continuous magnitudes.

(3) The third and final difference concerns the meaning of "a priori" and the demarcation line between empirical and a priori. On Meinong's view, all colors, including transcendent colors and the colors for which we have no words or concepts, stand in a priori relations that can be represented geometrically. By contrast, the a priori relations of color grammar are relations between a limited number of color words or concepts available in a given linguistic community. The fact that this orange is more similar to this red than to this blue is a mere empirical fact. What is a priori here is rather the fact that the word "orange" is normally used to denote a color that is more similar to a certain property normally called "red" than to a certain property normally called "blue". It may be false that this flower has a color that is more similar to blue than to red. But it is a priori impossible that this flower's orange color is more similar to blue than to red, because the word "orange" is normally used to denote a color that is more similar to red than to blue.

3 First Objection: Perceptual Contrast vs. Conceptual Distinction

At this point I would like to turn back to the three differences I just mentioned and, on the basis of them, suggest that the grammatical approach to color has some advantages over the geometrical approach.

The first difference was about Meinong's essentialism. Obviously, the idea that colors have a "nature" which is not merely conceptual and can be grasped through a priori intuition is a problematic idea. What does it mean for a color to have a "nature"? This nature must be something independent of the empirical data. On Meinong's view, the shape of the color solid is empirical, but the quasi-geometrical properties of colors cannot derive from, say, the structure of your visual field as it is described by empirical psychology. In some sense, it is undeniable that a priori relations among colors are independent of experience. For example: a red-blind person will have a completely different color solid than a normal person. This is an empirical fact which is due to her neurophysiological constitution. However, if a red-blind person tells you that red is not different from green, you will reply to her that she is *wrong* and that red is a priori known to be different from green. This is a priori knowledge: she has no experience of the color we call "red", and yet she ought to know that there is such-and-such a priori relations between the color we call "red" and other colors.

Now the question is, why are you right and why is she wrong? Do you have a priori intuitions that she is not able to have? This view seems unnatural and needlessly convoluted. Why not say more simply that a red-blind person can *talk* about colors that she is not able to *see*, that is, (incorrectly) use the concept of red? She can certainly talk about the red color, because normal persons have taught her that there is a color that is called "red" and is the color of ripe tomatoes. It is in no sense a matter of seeing. The red-blind person sees everything that needs to be seen. She does have color experiences, she sees ripe tomatoes with a certain color. The real point is that she is not able to say whether the color she sees is red or green. In other words, she is not able to talk correctly about colors, namely to make a correct use of the words "red" and "green".[24] So, if she tells you that red is not different from green, you will not say that she perceives red but does not see its real nature, but rather that she perceives a color which is not red,

24 This view is suggested by Wittgenstein 1977, p. 39: "In order to describe the phenomenon of red-green colour-blindness, I need only say what someone who is red-green colourblind cannot learn; but now in order to describe the 'phenomena of normal vision' I would have to enumerate the things we can do".

or that she wrongly uses the word "red" to denote a color that is not the color normally called "red".

In a nutshell: Her inability to perceive the contrast between this red and this green is an empirical fact to be explained by psychologists. Her inability to recognize the a priori difference between red and green is a linguistic matter: she is unable to recognize the a priori difference between red and green because her contingent inability to perceive the contrast between these reds and these greens prevents her from mastering the use of the words "red" and "green".

Suppose a friend comes to you and says this: "I just saw a magnificent flower in the garden. It was of a bright orange color. But that orange was very strange: it was not between red and yellow, but rather between blue and green." Of course, you will think that what she says makes no sense. It is a priori impossible for the orange color of the flower not to be between red and yellow. If the flower's color is not between red and yellow, then it must be another color than orange. But what will you reply to your friend? If you are a robust essentialist, then it seems that you must reply something like this: "Please go back in the garden and take a closer look at the flower so as to have a more accurate intuition of its orange color. Then you will see that the flower's orange color is not between blue and green, but rather between red and yellow." Such a reply sounds absurd. Of course no sensible person will conclude that your friend lacks a priori intuitions. You will more likely think that your friend is joking or that she does not master English language. Maybe you will lend her a dictionary and ask her to get a look not at the orange flower, but at the definition of the word "orange". Once again, your friend's inability to know that orange is between red and yellow is a linguistic matter.

There is another reason why you will not take seriously what your friend says about the flower's color. It is absurd to say that a given color is between two colors in some cases and not in other cases. Orange is between red and yellow in all possible cases and no empirical observation will ever show the existence of an orange color that is between blue and green. But here again this impossibility is better seen as a linguistic fact. Obviously, orange is in all possible cases between red and yellow just because the word "orange" is normally used to denote a color which is between the color usually called "red" and the color usually called "yellow". Thus, it is perfectly conceivable that in the future orange could be between green and blue, but this will then simply mean that some change in usage of color words has occurred.

All this suggests the Wittgensteinian conclusion that the a priori rules of the theory of color do not tell us anything about the nature of colors. The statement that orange is more similar to red than to blue is not a statement about orange color, it is just a rule according to which the word "orange" should be used only in cases where you experience a color that is more similar to red than to blue.[25]

4 Second Objection: Continuity

We come now to the second difference between color geometry and color grammar: the lines of the color solid are continua, while there appear to be no continua between color words or concepts. Meinong's geometrical approach has difficulties in this respect.

The problem is that for example the set of visible colors from red to yellow is not exactly a continuum as are geometrical lines. Two different colors can be too close to each other to be discriminable. The human eye can discriminate, say, up to ten million colors. This is much, and yet not enough to form a continuum. Accordingly, it seems that the geometrical representation misleads us into thinking that visible colors form a continuum.

Note that fine-grainedness raises different issues than the color solid limitation, even though both can somehow be explained by our psychological constitution. The problem, here, is not that some possible colors are invisible, but that some visible colors are different although indiscriminable.

Consider the case where two different colors on the red-yellow line are not discriminable. Are we to say that they are defined by the same coordinate triplet? On Meinong's view, if they are defined by the same coordinate triplet, then they are not two different colors, but one and the same color with one and the same "nature" – which contradicts our initial supposition. But if they are defined by different coordinate triplets and thus have a different "nature", then the whole theory becomes unintelligible. The part of color theory that deals with color solids studies colors as phenomenal properties – as phenomenal colors as opposed to light waves and the painter's pigments. In the phenomenal realm, a difference between two colors is called a "chromatic contrast". Now, we are asked to conceive of two colors that are different yet indistinguishable, that is, two colors that differ without contrasting. This seems to make no sense.

Maybe some would reply that for a red-blind person red and green are different colors that do not contrast. But this would be a poor argument. The red-blind

25 William Brenner has put the point very nicely in Brenner 1999, p. 122: "But it means nothing to speak of the grammar itself as true, or to talk of pointing to what makes it true. So if you say that color grammar is made true by 'the nature of the colors themselves', we respond with a puzzled 'The nature of *what?*' For is it not *grammar* that tells us what kind of object (e.g., color, shape, or sensation) something is?"

person does not see two different colors that do not contrast, but only one color. If you say that there really is two different colors and that the red-blind sees neither them nor the contrast between them, then you are talking about colors that are different and contrast in such-and-such optimal conditions, namely in day light, for a subject that is not red-blind, and the like.

A further, closely connected difficulty for Meinong's account relates to the intransitivity of color discrimination. It is obvious that, if every color is defined by a triplet of numerical values, then identity and difference between colors must be transitive relations. For identity and difference between numerical values are obviously transitive. If identity and difference between colors are not transitive, then it is false that necessarily two colors are (phenomenally) identical if, and only if, they correspond to one and the same coordinate triplet. However, it seems that the relations of phenomenal contrast and non-contrast between colors – unlike conceptual identity and difference – are non-transitive. A series of color patches can form a smooth transition from yellow to red such that each patch is indiscriminable in color from the one or two patches adjacent to it. Thus, the relation "looks the same color as" is not transitive. Likewise, the first patch of the series does not contrast in color with the second one, although both the first and the second patch do contrast with the last patch: the relation "contrasts in color with" is not transitive.[26]

Needless to say, since there is no continuum between color words or concepts, the above objections do not apply to the grammatical view.[27]

---

26 See, among many others, Goodman 1977; Wright 1975; Dummett 1975; Hardin 1988, p. 170 ff.; Rescher 2017, p. 60. Some authors have defended the view that such relations are transitive. See Jackson / Pinkerton 1973 and Graff 2001. A common response is that adjacent colors may be different even though they are indiscriminable. This suggestion, however, encounters the same difficulty as Meinong's view that two indiscriminable colors may have a different "nature" (see above). As Papineau (2015, p. 291) correctly points out: "The difficulty facing this line is obvious. In what sense can two samples have different conscious 'looks' if they are visually indiscriminable? Surely their indiscriminability means that they consciously look the same. The idea that these samples produce different conscious responses seems of dubious coherence. If the responses are consciously different, how can the subject be unaware of this?"

27 The question that arises is to what extent Wittgenstein *took seriously* the geometrical representations of the relations between colors in his *Philosophical Remarks*. Mulligan (1991, p. 85) observes very rightly that "the central difference between the two Austrian accounts of colour space is that Meinong, unlike Wittgenstein, takes seriously the notion of space that is appealed to in talking of tone space, colour space, etc.". Wittgenstein views the octahedron merely as a "*rough* representation of colour-space" (Wittgenstein 1977, p. 51).

5 Third Objection: Millions of Rules

The third difference mentioned above is that Meinong's color space includes transcendent colors and colors for which we have no words, while the a priori rules of color grammar apply to a limited number of color words or concepts.

An issue often raised about colors is that each color within the octahedron should stand in an enormous number of a priori relations to other colors. For instance, each color is different from each of the ten million or so other colors of the octahedron – let alone those that are outside the octahedron – and each difference can be expressed as a rule. Now, it is extremely implausible to say that color discrimination requires us somehow to know millions of rules. This applies also to conceptual differences between colors. We can ask whether the acquisition of the concept of green requires experiencing all visible green hues, and whether your concept of green is deficient in the converse case.[28]

Here again, it seems to me that a Wittgensteinian account is more plausible. Suppose we have only three words to denote all the color hues on the octahedron's line from pure red to pure yellow, namely "red", "yellow", and "orange". Each of the three words captures a third of the line, which means that all color hues within one of the three segments is denoted by one and only one of the three color words. So there are a huge range of color hues on the side of visual experience, while there are only three color words on the side of language.

As I understand it, Wittgenstein's view is very simple. It is that all the a priori relations dealt with in color theory are on the language side. Such relations hold not between colors merely as they are experienced, but between color words. Consider three different colors within the "orange" segment on the red-yellow line. Since the three colors are located in the "orange" segment, all should be called "orange". To distinguish them you will have to say "this orange", "that orange", "that other orange", etc. On Wittgenstein's view, the three experienced colors have no relations that can be called "a priori" in any intelligible sense of the term. This view is intuitive and plausible. The truth that this orange is between that orange and that other orange is not a necessary truth. It is no more than an observational truth: you are presented three colored patches, side by side, and observe that they stand in such-and-such relations to each other. It is certainly possible for this orange not to be between that orange and that other orange. By contrast, the truth that this orange is between red and yellow *is* a necessary truth, whose negation is self-contradictory. A person who denies it should not observe the orange patch more accurately, but consult a dictionary. Thus, it

28 Hardin 1988, p. 122.

may happen that somebody else, say, a Chinese person, has three different words to denote three orange hues you indiscriminately call "orange". In this case, the truth that Orange1 is between Orange2 and Orange3 is a necessary truth. But why is the situation different for the Chinese person? Obviously, the situation is different simply because the language is different!

At any rate, a Wittgensteinian treatment of the a priori has the effect of dramatically decreasing the number of a priori relations among colors so as to obtain quite a reasonable amount of rules – not only because the number of available color words is limited, but also because grammatical rules are such that a great number of them can be derived from other grammatical rules. For example, the conjunction of "A is between B and C" and "D is between A and B" entails that D is between B and C; the conjunction of "A is different from B" and "B is different from C" implies that A is different from C; the conjunction of "A is a shade of B" and "B is different from C" implies that "A is different from C"; etc.

6 Conclusion

Meinong's geometrical approach to the a priori relations between colors faces at least three sets of difficulties related to the following assumptions: colors possess a "nature" that can be grasped through intuition; they are separated from each other by continua in color space; there are an infinite number of a priori relations between colors. These difficulties do not arise for grammatical accounts. Therefore, grammatical accounts are more plausible in these three respects.

References

Barceló, Axel A. / Saab, Salma (2017), "The grammar of colours advanced in Wittgenstein's middle period", in: *Colours in the Development of Wittgenstein's Philosophy*, ed. by M. Silva, Palgrave Macmillan, p. 215–43.

Brenner, William H. (1999), *Wittgenstein's Philosophical Investigations*, Albany: State University of New York Press.

Carvalho, Marcelo (2017), "Colours, phenomenology and certainty: Wittgenstein's Remarks on Colours in the context of his later philosophy", in: *Colours in the Development of Wittgenstein's Philosophy,* ed. by M. Silva, Palgrave Macmillan, p. 317–337.

Crone, Robert A. (1999), *A History of Color: The Evolution of Theories of Lights and Color*, Dordrecht / Boston / London: Kluwer.

Dummett, Michael (1975), Wang's paradox, in: *Synthese* 30, p. 301–324.

Goodman, Nelson (1977), *The Structure of Appearance*, 3rd edition, Dordrecht: Reidel.

Graff, Delia (2001). "Phenomenal continua and the sorites", in: *Mind* 110, p. 905–935.

Hardin, C.L. (1988), *Color for Philosophers: Unweaving the Rainbow*, Indianapolis / Cambridge: Hackett.
Jackson, F.C. / Pinkerton, R.J. (1973), "On an argument against sensory items", in: *Mind* 82, p. 269–272.
Kuehni, Rolf G. / Schwarz, Andreas (2008), *Color Ordered: A Survey of Color Order Systems from Antiquity to the Present*, Oxford: Oxford University Press.
Meinong, Alexius (1903), "Bemerkungen über den Farbenkörper und das Mischungsgesetz", in: *Zeitschrift für Psychologie und Physiologie der Sinnesorgane*, Vol. 33.
Mulligan, Kevin (1991), "Colours, corners and complexity: Meinong and Wittgenstein on some internal relations", in: *Existence and Explanation: Essays in Honor of Karel Lambert*, ed. by B.C. van Fraassen, B. Skyrms and W. Spohn, Dordrecht: Kluwer, p. 77–101.
Papineau, David (2015), "Can we really see a million colours?", in: *Phenomenal Qualities: Sense, Perception, and Consciousness*, ed. by P. Coates & S. Coleman, Oxford: Oxford University Press.
Park, Byong-Chul (1998), *Phenomenological Aspects of Wittgenstein's Philosophy*, Dordrecht: Springer.
Rescher, Nicholas (2017), *Epistemic Principles: A Primer for the Theory of Knowledge*, New York: Peter Lang.
Rollinger, Robin (2008), "Meinong on the objects of sensation", in: *Austrian Phenomenology: Brentano, Husserl, Meinong, and Others on Mind and Object*, Frankfurt am Main: Ontos Verlag, p. 109–137.
Seron, Denis (2015), "Les couleurs, la phénoménologie et la grammaire", in: *Philosophiques* 42/2, p. 385–389.
Seron, Denis (2016), "Des mots, des couleurs et des octaèdres", in: *Bulletin d'Analyse Phénoménologique* XII (2), p. 139–169.
Wittgenstein, Ludwig (1929), "Some remarks on logical form", in: *Proceedings of the Aristotelian Society*, Supplementary Volumes, 9, p. 162–171.
Wittgenstein, Ludwig (1963), *Tractatus Logico-Philosophicus*, Engl. transl. by D.F. Pears & B.F. McGuinness, London: Routledge & Kegan Paul.
Wittgenstein, Ludwig (1975), *Philosophical Remarks*, Oxford: Blackwell.
Wittgenstein, Ludwig (1977), *Remarks on Colour*, Oxford: Blackwell.
Wright, Crispin (1975), "On the coherence of vague predicates", in: *Synthese* 30, p. 325–65.

Carlo Ierna

Thinking the Impossible: The *Gestalt* of a Round Square

Abstract: In this article I connect two concepts that played central roles in the School of Meinong: the notion of impossible objects and that of Gestalt. Ehrenfels claims that Widerspruch or incompatibility would be a temporal Gestalt quality, specifically the trying and failing to build an intuitive presentation. Where, when, and how does this process break down exactly? Meinong's Graz students developed a more detailed production theory for the presentation of Gestalten (Vorstellungsproduktion) which can help to determine how the failure to produce one Gestalt can generate another, which moreover is typical of impossible objects. There are three contributions to Meinong's 1904 volume Untersuchungen zur Gegenstandstheorie und Psychologie, the ones by Mally, Benussi, and Ameseder, that provide helpful analyses to clarify the incompatibility Gestalt. These suggest three options to refine the analysis of the Gestalt of impossible objects. The conclusion is that the failure of the Vorstellungsproduktion itself is not enough, the incompatibility relation between the foundations must also play a role.

1 Ehrenfels' Notion of *Gestalt*

In his development of the notion of *Gestalt* Ehrenfels was in part inspired by Mach's *Beiträge zur Analyse der Empfindungen*[1], as he himself points out, but we can find the notion of *Gestalt*, or at least some embryonic concept of it, already previously in the School of Brentano, e. g. in Brentano himself,[2] in Stumpf's discussion of fusion,[3] and in Husserl,[4] who was inspired much by the same sources and context as Ehrenfels.[5] In the present article, however, I'd like to focus on how the concept was then later adopted and adapted in the School of Meinong.

1 Mach 1886, see, e. g., p. 43, 47, 128. Cf. Mulligan / Smith 1988 for a more extensive discussion.
2 E. g., Brentano 1884/85, p. 58 f.
3 But also in Stumpf 1883, p. 115.
4 Husserl 2005.
5 See Ierna 2005; 2009.

Carlo Ierna, Radboud Universiteit Nijmegen
carlo.ierna@gmail.com

https://doi.org/10.1515/9783110664850-004

Indeed, looking at this context, we can see how the notion of *Gestalt* originated through a dialogue across a series of publications between Ehrenfels and his teacher Meinong. Starting with the theory of relations in Meinong's 1882 *Hume-Studien* II, through Ehrenfels' 1885 dissertation on *Relations of Magnitude and Numbers*,[6] which contains early seeds of the notion, and of course his titular 1890 article on *Gestalt* qualities, then Meinong's 1891 response in "*Zur Psychologie der Komplexionen und Relationen*", we see the concepts of *Gestalt* and founded object slowly crystallize.

It is always easy to explain a philosophical concept by using the paradigmatic examples, e. g. in the case of *Gestalt*, a geometrical figure or a melody. These can be transformed or transposed, in such a way that all their constitutive elements change their position, but their constitutive relations remain constant. Hence, we can have "the same" melody in different keys, etc. However, things become much more difficult, but also much more interesting, when looking at unusual and extreme cases.

Ehrenfels makes a fundamental distinction in temporal (*Zeitliche*) and non-temporal (*Unzeitliche*) *Gestalten*. With temporal *Gestalt* qualities "at most one element can be given in perceptual presentation", with non-temporal *Gestalt* qualities the "foundation can be given completely in perceptual presentation".[7] However, *Gestalten* are not limited to a single sense, but can combine multiple senses:

> The intimate fusion of touch, temperature and sometimes also taste and smell sensations into a single unified total impression raises the question of whether we do not have here Gestalt qualities which are built up upon a foundation belonging to *several sensory regions*. [...] we must also accept the possibility of Gestalt qualities comprehending complexes of elements of different categories.[8]

Temporal Gestalt qualities are generated by "every change of any kind in a presentational complex, providing it proceeds in some definite direction", if such

6 Ehrenfels 1885, published as Ehrenfels 2017, also see Ierna 2017. Consider e. g. Ehrenfels' conclusion: "Therefore we will find complexes in the most disparate domains of what we can present, which all the same display a clearly observable similarity in their relations, such that even the most incompatible things can obtain unity and harmony through this relation" ("Darum werden sich auf den verschiedensten Gebieten des Vorstellbaren Complexe auffinden lassen, welche gleichwol in ihren Verhältnissen eine deutlich wahrzunehmende Ähnlichkeit zeigen, so daß durch die Relation auch dem scheinbar Unvereinbarsten Einheit und Harmonie verliehen werden kann").

7 Ehrenfels 1890/1988, p. 263–64/94.

8 Ehrenfels 1890/1988, p. 267–68/97.

change "reveals a unified character, so that we have (or could have) a name to designate it".

> We have to make do with a few words for isolated examples from the whole range of possible colour changes (e. g. blushing, blanching, darkening, glowing, etc.). [...] Beside the musical temporal qualities already considered we have to recognize also such noise or resonance Gestalten as, say, thundering, exploding, rustling, splashing, etc.[9]

However, not only change, but also persistence (*Dauer*), such as the sustaining of a note,[10] can give rise to a peculiar *Gestalt* quality. In this way we see that for Ehrenfels practically everything we ordinarily experience involves *Gestalten* of some kind: "the larger part of both our everyday and our scientific vocabulary designates *Gestalt* qualities": all verbs (with few exceptions such as the auxiliaries), all nouns, and all adjectives that refer to "more than a single presentational element".[11]

Gestalt qualities are not limited to outer perception, but also occur in inner perception, where the growing or diminishing of desires, fears, expectations, etc. are compared to a crescendo or diminuendo in music and they arise in the case of actions "expressed by means of common nouns (kindness, service, rivalry, marriage, theft, war, etc.) or verbs (entreat, complain, help, rob, avenge, etc.)".[12] Hence, we have to acknowledge *Gestalten* based on a "union of the physical and the psychical", including the process of fulfilment of unintuitive presentations by intuitive ones.

2 The *Gestalt* of Incompatibility

Connected to this case, we finally get to the problematic case that I'd like to examine in more detail: *Widerspruch* or incompatibility. Ehrenfels introduces it as a "phenomenon of the highest importance which can be understood as a *Gestalt* quality". Ehrenfels then immediately dives into a discussion of intuitive and unintuitive presentations in Meinong. By comparing cases in which an intuitive presentation of compatible contents is possible with cases in which an intuitive presentation is impossible, Ehrenfels aims to draw out a general characteristic that is common to all evident negative judgements about the

9 Ehrenfels 1890/1988, p. 269, 272/98, 100.
10 Ehrenfels 1890/1988, p. 273/101.
11 Ehrenfels 1890/1988, p. 281-282/108.
12 Ehrenfels 1890/1988, p. 281/107.

possibility of combining incompatible contents into an intuitive presentation, i.e. a general characteristic of impossible presentations. After clarifying what he means by intuitive and unintuitive through his analysis of Meinong, Ehrenfels points out that we can also present compatible contents in an unintuitive, empty fashion, since it is in the most cases more convenient to do so. Indeed, only thanks to the fact that we can form proper as well as improper presentations of the same objects, we can have the possibility of passing from an unintuitive presentation to an intuitive presentation of them:

> Anyone confronted with, say, a complicated description of a work of architecture will first of all form a merely indirect presentation of it, which will then be rounded out by gradual execution or fulfilment of the various merely intended components, to yield an intuitive total picture. But this process of formation of the intuitive presentation directly from the indirect presentation is something that happens, a process of change, which serves as the foundation for a specific temporal Gestalt quality.[13]

According to Ehrenfels "it requires only one step more to provide an explanation of incompatibility." Let us look at how Ehrenfels describes the process in detail:

> For if one now attempts to bring to the state of executed combination an empty intention of incompatible determinations such as round and square, [...] then the process thereby set in motion is initially wholly analogous to this earlier case, until suddenly there comes into play a certain moment, not more precisely describable, in which (speaking metaphorically) the determinations resist becoming unified, as if one were to try to force two bodies into a container that had been constructed to take only one of them. [...] The formation-process now falters at the half-way point in an unmistakable way – not precisely characterizable by means of comparisons – which we call 'incompatibility'.[14]

Hence, Ehrenfels concludes that incompatibility, i.e. the failure to attain an intuitive presentation of impossible objects, is a temporal *Gestalt* quality "which determines the conviction of the non-existence of those objects in whose attempted formation in intuitive presentation it has arisen." This moreover prompts him to claim, without elaborating, that this specific *Gestalt* quality is really at the heart of the notion of evidence, since all affirmative evident judgements could be based on the evident incompatibility of their negative counterparts: i.e. possibility is the negation of impossibility.

---

13 Ehrenfels 1890/1988, p. 276-277/104.

14 Ehrenfels 1890/1988, p. 277/104.

3 Meinong and Ehrenfels on Incompatibility

However, this account of the experience of incompatibility and its peculiar *Gestalt* quality has some serious flaws, which Meinong pointed out in his 1891 reaction to Ehrenfels' article. The heaviest objection would seem that Ehrenfels neglects the fundamental distinction between the two layers of every intentional act, i.e. the primary act directed at the contents themselves and the secondary act reflectively directed at the primary act and contents. The primary acts are directed at the contents, but in the reflective acts of internal perception I am aware of them as presentations of contents. Meinong hence argues against Ehrenfels that:

> Compelling though the description of these experiences might be, the conception of the Author [i.e., Ehrenfels] seems to leave no option open to reject the objection that according to it actually "round" and "square" are not the founding contents, but rather the internally perceived contents "presentation of round" and "presentation of square".[15]

Of course, these might be incompatible too, but that would be a quite different claim. The impossibility of an intuitively presented round square is different from the impossibility of a round square. A geometrical figure cannot be at the same time round and square, i.e. roundness and squareness cannot inhere in an object at the same place and time, however, I can perfectly well form the presentation "round square" unintuitively and symbolically: I just did so linguistically. Meinong correctly points out that on the level of presentations the additional claim needs to be made that roundness and squareness are incompatible in an intuitive presentation.

However, Meinong does rightfully acknowledge that Ehrenfels is in fact trying to improve on Meinong's own earlier attempts at a definition of intuitability through the denial of incompatibility in "*Phantasie-Vorstellungen und Phantasie*":[16]

> We often have the experience that the unintuitive presentational connection is unaffected by contradictions, i.e. that these, if they occur, only become apparent when we try to transition to an intuitive presentation. Naturally it is not the factual failure of the attempt and the experience based thereupon, but the evidence resulting from these circumstances, that justifies the judgment of incompatibility. Therefore, the preceding investigations legitimate the following statement: "a complex presentation is intuitive in so far as it is free from incompatibility in all directions".[17]

---

15 Meinong 1891, p. 262.

16 Meinong 1889, p. 213.

17 "Es ist eine oft gemachte Erfahrung, daß die unanschauliche Vorstellungsverbindung gegen Widersprüche gleichsam unempfindlich ist, d.h. diese, wenn vorhanden, erst zutage

Indeed, in a footnote in his 1890 article Ehrenfels chides Meinong about having provided this definition, which Ehrenfels considers viciously circular:

> Meinong's attempted definition reads as follows: "A complex presentation is intuitive in so far as it is free of incompatibility". It seems to me to be unacceptable, since the concept of incompatibility can itself only be understood in terms of the distinction between empty and executed presentational combinations, which Meinong identifies with unintuitive and intuitive, respectively. Thus the definition presupposes precisely the concept which it sets out to define.[18]

Although Meinong now in 1891 correctly points out that nevertheless any contributions from the perspective of internal perception are irrelevant to the incompatibility of the contents, I find this a very unsatisfying reply that does not get to the meat of the matter. If the typical *Gestalt* of incompatibility arises precisely through the failure of production and if we need such a *Gestalt* to actually define intuitability, then we need a more sophisticated production theory that traces the various phases of the formation-process Ehrenfels describes and points out more precisely what the "resistance" is. Where, when, and how does the *Vorstellungsproduktion* break down?

The way in which Ehrenfels describes the process of attempting to unify incompatible contents into a whole, speaking of "formation-process" does seem to suggest some kind of production theory. This issue whether *Gestalt* qualities are the result of a mental act or are passively given, would soon lead to a profound divide between the Berlin and Graz schools of *Gestalt* psychology: do we actively construct *Gestalten* or do we merely notice them? Despite the terminology used in the description of the experience of incompatibility, Ehrenfels did not yet develop a fully clear position on this in his 1890 article. Indeed, in other parts of the text he seems rather opposed to something like a production theory. Ehrenfels admits that even though we can actively switch between different *Gestalten* by changing the focus of our attention, as is the case with the Necker cube, we do not actually produce the *Gestalt*, but merely notice it.

Meinong and Ehrenfels do not provide a fully satisfactory account of the incompatibility *Gestalt* in the case of impossible objects. Can we find a satisfactory

treten, wenn der Übergang zum anschaulich Vorgestellten versucht wird. Natürlich ist es aber nicht etwa das tatsächliche Mißglücken das Versuches und die darauf gestützte Erfahrung, sondern die unter diesen Voraussetzungen eintretende Evidenz, was das Unverträglichkeitsurteil begründet. Berechtigen uns sonach die vorstehenden Untersuchungen zu dem Satze 'Anschaulich ist eine komplexe Vorstellung, sofern sie nach jeder Richtung frei von Unverträglichkeit ist'".

18 Ehrenfels 1890/1988, p. 276 n. 1/117 n. 4.

account of the *Gestalt* of incompatibility in the subsequent work of Meinong's Graz School?

4 Incompatibility in the Graz School

The School of Meinong went on to develop a full-fledged production theory. Higher order objects are actively produced by a mental effort, operating on the lower order constituents. The *Gestalt* as a whole is actively made out of its founding contents. It is only through a specific activity (referred to as "*Gestaltbildung*" or as a form of "*Vorstellungsproduktion*") that we can add the non-sensory higher order *Gestalt* on top of the primary contents of sensation. Instead of considering the *Gestalt* as a quality given at once together with its foundations, the members of the Graz school considered it to be a separate object. They did not speak of "*Gestaltqualitäten*", but of "*Gestaltgegenstände*" and called the foundations *inferiora* and the founded *Gestalt* a *superius*.[19] The foundations, i.e. sensory contents, were considered to be "real", but the founded object, i.e. the *Gestalt*, was considered to be "ideal".[20] Moreover, this relation of "founding" between the foundation and the founded was explicitly distinguished from the process of causation.[21]

Now, doesn't the incompatibility *Gestalt* present a problem for this kind of theory? How can the failure to produce one *Gestalt* generate another, which moreover would be typical of impossible objects? There are three contributions to Meinong's 1904 volume *Untersuchungen zur Gegenstandstheorie und Psychologie*, specifically the ones by Mally, Benussi, and Ameseder, that provide helpful analyses to clarify the incompatibility *Gestalt*, though none address it directly.

In "*Zur Gegenstandstheorie des Messens*", Mally extensively discusses impossible objects such as the round square, but mainly focusing on issues that are peripheral to our concern here. Mally points out that despite the independence of *Sein* and *Sosein*, objects such as the round square, cannot be precisely because of their incompatible properties. Objects with incompatible properties cannot be

19 Meinong (1891, p. 253) introduced the new terminology of "founding" and "founded" contents, because he considered "shape" and "quality" equivocal. Ehrenfels (1891, p. 293) himself then also took over this nomenclature. Also see Ameseder 1904, p. 110 ff.

20 Meinong (1899) proposes two different distinctions between real and ideal in § 6 and § 7 of his *Über Gegenstände höherer Ordnung*. In § 7 ("*Erfahrungsgegenstände und fundierte Gegenstände*") he endorses the distinction as given here between real and ideal based on their perceivability: real objects can be perceived, ideal objects cannot. Foundation serves an analogous role for ideal objets, as perception for real objects (Meinong 1899, p. 203).

21 Benussi 1904, p. 383. This paragraph is partially based on Ierna 2018.

and are called impossible objects. Possible objects are all those with non-incompatible properties. Mally further qualifies that what is impossible is the being of a round square, and not the being round or the being square of a round square, which are objective in their own right. Nevertheless, since the round square cannot be, neither do its properties, so there is no roundness and squareness of the round square. Quite obviously then, their combination "square roundness" cannot be at all.

> It *subsists* that the round square *is round and square* (i.e. its Being-such); however, the *quality of the round shape* and the *quality of the square shape* do not subsist in this object, and the encompassing quality of the impossible object, which we could call "square-roundness", does not subsist at all.[22]

Mally argues that impossible objects do not really have their incompatible properties, since qualities such as shape can only inhere in real objects.

> Qualities *in* objects are e. g. color and shape. These are properties that are not objectives, but objects in the strict sense. [...] Color can only *exist* in an *existing* colored object, shape only in an *existing* shaped object.[23]

What we are looking for, however, is not the ontological status of impossible objects and their properties, but the specific *Gestalt* of incompatibility itself. This would be a higher order *Gestalt*, based on the relation between the incompatible properties. Mally lists as relational *Gestalten* "similarity, equality, difference, compatibility" ("*Ähnlichkeit, Gleichheit, Verschiedenheit, Verträglichkeit*"). Can we add "incompatibility" ("*Unverträglichkeit*") to the list? The answer is not obvious, but Mally's text seems to suggest it cannot. After all, the round square doesn't actually have the qualities "round" and "square", hence they cannot stand in any real relations. Can they stand in ideal relations? This too isn't fully clear, but Mally uses a similar example which also suggests that "incompatibility" cannot be understood as a relation:

---

22 Mally 1904, p. 129: "Es *besteht* zwar, daß das runde Viereck *rund und viereckig ist* (d. h. sein Sosein); allein die *Qualität der Kreisgestalt* und die *Qualität der Vierecksgestalt* besteht an diesem Gegenstande nicht, und die Gesamtqualität des unmöglichen Gegenstandes, die man etwa "Viereckig-Rundheit" nennen könnte, besteht überhaupt nicht."

23 Mally 1904, p. 141 f.: "Qualitäten *an* Gegenständen sind z. B. Farbe und Gestalt. Sie sind Eigenschaften, die keine Objektive, sondern Objekte im engeren Sinne sind. [...] Farbe kann nur an einem *seienden* farbigen Gegenstande sein, Gestalt nur an einem *seienden* gestalteten Gegenstande".

> An ideal relation does also not subsist if the essentially coinciding Being-such is not impossible, but contradictory: the objective that the A and B which are (in the same respect) equal and unequal, are equal and unequal, subsists, but there is no relation with the combined relatum "equal and unequal" as a quality between A and B, because it is at all impossible.[24]

Mally indeed also describes the case of failure when trying to form a complex. The passage from an unintuitive to an intuitive presentation of a complex he calls synthesis. This synthesis can fail, but Mally only considers the case of contingent psychological limitations, e. g. when we try to intuitively present large numbers. We can intuitively conceive of the "trinity" complex when looking at three objects, but fail to form an intuitive complex when confronted with 12 or more objects. However, Mally does not suggest that the failure of the synthesis is itself again a different kind of *Gestalt* or would have certain peculiar characteristics of its own.

The failure to form a specific *Gestalt* is addressed more in detail in the contribution of Benussi "*Zur Psychologie des Gestalterfassens*". Benussi concentrates more in detail on the way in which we operate on the foundations or inferiora to produce the higher order *Gestalt* or superius. The simple presence of the inferiora alone is not enough on its own to produce a *Gestalt*. Various circumstances can aid or hinder the production of the *Gestalt*. Benussi investigated in great detail how the Müller-Lyer illusion is affected by various changes in color, brightness, contrast, etc. of the elements as well as the attitude of the test subject. Benussi clarifies that there is a fundamental distinction between the causally effective presentations of the inferiora and the ideal nature of the produced *Gestalt*. Hence, when we fall prey to an illusion we can ask whether the error in judgement arises in an inadequate presentation of the inferiora or an inadequate production of the superius. Benussi argues that the Müller-Lyer illusion ultimately is not due to an inadequacy of the sensations, but of the process of production. Might this approach contribute to explain the failure to form a *Gestalt* out of incompatible elements as well, as in the case of the round square? This does not seem likely. Benussi points out that in the case of the variations on the Müller-Lyer illusion we can observe degrees of inadequacy or ambiguities between two possible *Gestalten* based on the same inferiora. This is inapplicable to round squares which are in principle impossible to present intuitively. The round square doesn't fail to be

---

24 Mally 1904, p. 146: "Eine Idealrelation besteht auch nicht, wenn das wesentlich koinzidierende Sosein nicht unmöglich, aber widersprechend ist: das Objektiv, daß A und B die (in derselben Hinsicht) gleich und ungleich sind, eben gleich und ungleich sind, besteht; es besteht aber keine Relation mit dem zusammengesetzten Relat "gleich und ungleich" als eine Qualität zwischen A und B, – denn sie ist überhaupt unmöglich".

produced by degrees nor is it an ambiguous figure that appears as square or as circle depending on the circumstances. Hence, the failure to produce the illusion, e. g. by different colorations of the central lines whose lengths are to be compared, is of a completely different kind than the failure to intuitively present a square circle, which means there should be two different processes of production which are responsible for the differences in the inadequate presentation in the case of the Müller-Lyer illusion and the impossible presentation in the case of the square circle.

So let us take a closer look at the process of production, as analyzed in detail in Ameseder's contribution "*Über Vorstellungsproduktion*". Ameseder begins with an outline of the structures of dependency in founded objects. Founded objects are something ideal and dependent on their foundations while the foundations are independent from the *Gestalt*. Hence, we cannot present a relation of difference without two foundations that are different. Yet, we can present the foundations independently from their difference. We might now suppose that the presentation of the founded content is related to the presentations of the foundations in the same way as the founded content to the foundations. This quite appealing hypothesis, however, is not the case. Like Meinong, also Ameseder points out that there is a fundamental difference between e. g. a relation between the elementary foundations on the one hand and a relation between the presentations of the elementary foundations. Moreover, presentations of founded contents are a special case:

> The difference of the presentations A and B (v1) however, is not identical with the presentation (V) of the difference of a and b (v). This difference-presentation therefore cannot be founded, given that everything founded, and hence also something founded by these presentations, indeed cannot be something real, but a presentation would possibly have to be something real. [...] Hence, the presentations of founded objects are not founded. Doubtlessly they do build on the presentations of the inferiora, so this kind of constructing must be different from founding.[25]

So in the case of the round square we would have to distinguish not only the roundness and the squareness of the round square from the presentation of the

25 Ameseder 1904, p. 487: "Die Verschiedenheit der Vorstellungen A und B (v1) ist aber natürlich nicht identisch mit der Vorstellung (V) der Verschiedenheit von a und b (v). Diese Verschiedenheits-vorstellung kann somit nicht fundiert sein, da alles Fundierte, also auch etwas durch diese Vorstellungen Fundiertes, eben nichts Wirkliches sein kann; eine Vorstellung aber müßte wirklich sein können. [...] Fundiert sind also die Vorstellungen fundierter Gegenstande nicht. Daß sie sich gleichwohl auf die Inferioravorstellungen aufbauen ist zweifellos; somit muß die Art dieses Aufbauens eine andere sein als die Fundierung." (Cf. previous and next footnote.)

roundness and the presentation of the squareness, but also the way the round square is founded on its roundness and squareness from the way that the presentation of the round square is based on the presentation of its foundations. This is precisely what is then termed as *Vorstellungsproduktion*. It is not the round circle that is produced, but the presentation of the round circle. It is crucial now to find out how the failure of producing an intuitive presentation of the round square is related to the foundation of the round square. As Ameseder points out in the case of the relation of difference: "To the contrary, we do not produce a difference, but the presentation of it; the difference is *grasped through* the production".[26]

In the case of the relation of incompatibility this would mean that the relation itself is founded by the contents and then grasped through the production of its presentation. The contents themselves, roundness and squareness can both be presented and the relation of incompatibility obtains between them. However, the relation of incompatibility is not what we are looking for, we are looking for the specific *Gestalt* that is occasioned when the production of the presentation of a founded object fails, because the founded object itself is impossible. We can only have indirect, unintuitive presentations of the round square, the attempt to have a direct, intuitive presentation fails. Ameseder addresses the case of indirect presentations, treating them as pre-given goals toward which the *Vorstellungsproduktion* then strives. The indirect presentation helps to select the correct production process, such as e. g. comparing, etc. Ameseder argues that to every kind of *superius* corresponds a specific kind of production process. So we would have production processes for similarity, for difference, etc. Yet this does not elucidate the case where the *superius* is an impossible object. I can very well imagine a series of production processes in the course of a successful geometrical construction, of drawing lines, comparing distances, grasping figures, etc. but what happens when the indirect presentation, the target that is to be constructed is impossible? There seems to be no fitting production process to generate a round square. In this case the *Vorstellungsproduktion* would not break down at all, but never even get off the ground. This might seem the obvious option to take, but outside of the obvious cases of the round square where we already know that its foundations are incompatible, we should be able to account for cases where this is not obvious form the start. We might not have a production process for the presentation of impossible objects, but we can certainly find ourselves involved in the attempt to produce a presentation of something we do not yet know to be

26 Ameseder 1904, p. 488: "Dagegen produziert man nicht eine Verschiedenheit, sondern die Vorstellung einer solchen; die Verschiedenheit erfaßt man durch Produktion".

impossible. Imagine you have just discovered how to trisect a line using only straightedge and compass and now cheerfully try to trisect an angle.

5 Concluding Remarks

This is precisely the kind of situation Ehrenfels describes in his article. Not like Mally a failure due to the overflowing of our working memory when dealing with too many elements, not like Benussi degrees to which the *Gestalt* of an optical illusion become more or less prominent, not like Ameseder the lack of an appropriate production process. Ehrenfels points to a very specific *Gestalt* that occurs only in the case of impossible objects with incompatible foundations. Despite all the corrections and details developed in the theories and experiments of Meinong and the Graz school, there seems to be no clear account of why this would occur nor an argument why Ehrenfels would have been wrong in his description. Again, the point here is not to discuss the ontological status or the *Außersein* of impossible objects, but the specific *Gestalt* of the failure of the process of production. If we want to claim that the reason why the process fails, is that the foundations are incompatible and the target object that is presented indirectly, the round square, is impossible, then this calls for a clarification of the relation between the foundation and the process of production in the cases where one succeeds and the other fails, specifically where it fails for a priori reasons, not a posteriori reasons that have to do with the contingent limits of attention, since these imply at least a partial success.

What are our options? 1) Given two incompatible foundations, without regarding their incompatibility in particular, we can form an unintuitive presentation of the founded object and then look for an appropriate production process to apply to them. This search fails, there is no appropriate process and the production never gets off the ground. 2) We somehow start up the production process, but this process fails, just like in the case when we surpass the *Enge des Bewusstseins*, the narrow mind of man, the only difference is that no partial or intermediate stages of intuitability were reached. 3) The production process starts, fails, and this failure itself, due to, as Ehrenfels says, the "unmistakable way – not precisely characterizable by means of comparisons" that occurs only in the case of incompatible foundations, actually kickstarts a second involuntary production process which generates the typical *Gestalt* we are looking for.

All three options have unappealing aspects. We want to be able to consider round squares and impossible objects in general as higher order, founded objects. We also, following Meinong's analysis that goes back all the way to the

Hume-Studien II, accept that incompatibility is the fundamental notion that determines the unintuitability of impossible objects. Yet, both Meinong and Ehrenfels concur that beyond the founded object itself and beyond the unintuitive presentation of objects with incompatible properties, one can and does make an attempt at forming, at producing an intuitive presentation. This attempt at *Vorstellungsproduktion* then breaks down. All things considered the third option is the one that stays closest to the description offered by Ehrenfels. After all, *Vorstellungsproduktion* is itself a process giving rise to a *Gestalt*, and the failure of a *Vorstellungsproduktion*, is likewise the foundation for a temporal *Gestalt*, which can be presented. So we would have the presentation of the *Gestalt*, resulting from the specific kind of *Vorstellungsproduktion* that is applicable to the failure of a *Vorstellungsproduktion* due to the incompatibility of the foundations.

However, we're not there yet. *Vorstellungsproduktion* may break down in other cases and due to other reasons too. What makes the failure to form an intuitive presentation of a square circle different from the failure to intuitively present, say, a circle with a thousand sides? As Meinong pointed out, Ehrenfels' description only focuses on the incompatibility between the presentations of the foundations. What distinguishes the two cases is the incompatibility relation between the foundations themselves, so we have to allow the incompatibility relation between the foundations to play a role. Hence, the typical *Gestalt* that Ehrenfels identifies cannot be founded exclusively on the failure of the *Vorstellungsproduktion* itself, but must also involve the incompatibility relation between the foundations. In other words, we can address Meinong's objection by claiming that the specific kind of incompatibility we experience between the presented foundations can only occur if there actually also is an incompatibility between the foundations themselves. Hypothetically, the pathway could be as follows. The typical way in which the presentation of the foundations resists an intuitable unification involuntarily draws our attention to the incompatibility relation between the foundations, which can but does not have to be noticed, and together with the breaking down of the original *Vorstellungsproduktion* involuntarily occasions the second process of *Vorstellungsproduktion* which characterizes the attempt to think the impossible.

References

Ameseder, Rudolf (1904), "Beiträge zur Grundlegung der Gegenstandstheorie", in: Meinong (1904), p. 51–120.

Benussi, Vittorio (1904), "Zur Psychologie des Gestalterfassens (Die Müller-Lyersche Figur)", in: Meinong (1904), p. 303–448.

Brentano, Franz (1884/85), *Ms. Y 3: Die elementare Logik und die in ihr nötigen Reformen II*, unpublished lecture notes from the Husserl-Archives Leuven.
Ehrenfels, Christian von (1885), *Über Größenrelationen und Zahlen. Eine psychologische Studie*, Dissertation, Universität Graz.
Ehrenfels, Christian von (1890), "Über 'Gestaltqualitäten'", in: *Vierteljahrsschrift für wissenschaftliche Philosophie* 14, p. 249–292, reprinted in: Fabian 1988, p. 128–167.
Ehrenfels, Christian von (1891), "Zur Philosophie der Mathematik", in: *Vierteljahrsschrift für wissenschaftliche Philosophie* 15, p. 285–347.
Ehrenfels, Christian von (1988), "On 'Gestalt Qualities'", Engl. transl. by B. Smith, in: Smith, Barry ed., *Foundations of Gestalt Theory*, Munich / Vienna: Philosophia Verlag, p. 82–117.
Ehrenfels, Christian von (2017), "*Über Größenrelationen und Zahlen. Eine psychologische Studie*", Ierna, Carlo ed., in: *Meinong Studies* 8, p. 185–234.
Ierna, Carlo (2005), "The Beginnings of Husserl's Philosophy (Part 1: From *Über den Begriff der Zahl* to *Philosophie der Arithmetik*)" in: *The New Yearbook for Phenomenology and Phenomenological Philosophy* 5, p. 1–56.
Ierna, Carlo (2009), "Husserl et Stumpf sur la Gestalt et la fusion", in: *Philosophiques* 36/2, p. 489–510.
Ierna, Carlo (2017), "On Ehrenfels' Dissertation", in: *Meinong Studies* 8, p. 163–184.
Ierna, Carlo (2018), "Christian von Ehrenfels on the Mind and its Metaphysics", in: Lapointe, Sandra ed., *History of Philosophy of Mind in the 19th Century*, *History of Philosophy of Mind*, vol. 5, London: Routledge, ch. 12.
Husserl, Edmund (2005), "Lecture on the Concept of Number (WS 1889/1890)", ed. and transl. by C. Ierna, in: *The New Yearbook for Phenomenology and Phenomenological Philosophy* 5, p. 278–309.
Mach, Ernst (1886), *Beiträge zur Analyse der Empfindungen*, Jena: Fischer.
Mally, Ernst (1904), "*Zur Gegenstandstheorie des Messens*", in: Meinong (1904), p. 121–262.
Meinong, Alexius (1882), "Hume Studien II: Zur Relationstheorie", in: *Sitzungsberichte der philosophisch-historischen Klasse der Kaiserlichen Akademie der Wissenschaften in Wien* CI:II, p. 573–752.
Meinong, Alexius (1889), "Phantasie-Vorstellungen und Phantasie", in: *Zeitschrift für Philosophie und philosophische Kritik* 95, p. 161–244.
Meinong, Alexius (1891), "Zur Psychologie der Komplexionen und Relationen" in: *Zeitschrift für Psychologie und Physiologie der Sinnesorgane* 2, p. 245–265.
Meinong, Alexius (1899), "Über Gegenstände höherer Ordnung und deren Verhältniß zur inneren Wahrnehmung", in: *Zeitschrift für Psychologie und Physiologie der Sinnesorgane* 21, p. 182–272.
Meinong, Alexius (1904, ed.), *Untersuchungen zur Gegenstandstheorie und Psychologie*, Leipzig: Barth.
Mulligan, Kevin / Smith, Barry (1988), "Mach and Ehrenfels: The Foundations of Gestalt theory", in: Smith (1988), p. 124–157.
Smith, Barry ed. (1988), *Foundations of Gestalt Theory*, Munich / Vienna: Philosophia Verlag.
Stumpf, Carl (1883), *Tonpsychologie*, vol. 1, Leipzig: Hirzel.

Philipp Berghofer

What Role Can Experimental Psychology Play in Philosophical Investigations: Meinong and Benussi as Precursors of Experimental Philosophy?

Abstract: Experimental philosophy is often considered a novel and exciting movement in philosophy that paves the way for new ways of interdisciplinary research and offers an unprecedented methodology that allows gaining unexpected results that could revolutionize philosophical research. In this paper, however, we shall see that the methodology of experimental philosophy has been well-known and even practiced at Meinong's Graz school of object theory and experimental psychology. This is particularly true for the works of Vittorio Benussi that can be regarded as an experimental confirmation of some of Husserl's central theses. We shall also shed light on some shortcomings of how experimental philosophy is currently practiced and argue for bringing an intuition-free experimental philosophy into focus. In the final section, we will see that Benussi's experimental research qualifies as such an intuition-free experimental philosophy.

1 What is Experimental Philosophy?

Experimental philosophy (X-Phi) is often considered a novel and exciting movement in philosophy that paves the way for new ways of interdisciplinary research and offers an unprecedented methodology that allows gaining unexpected results that could revolutionize philosophical research. Interestingly enough, however, there is no agreement on the precise definition of X-Phi. Furthermore, after nearly two decades of practicing X-Phi, it remains unclear what the revolution is supposed to consist in or what exactly we have learned that was formerly unexpected but is uncontroversial now. To be sure, there is no doubt that X-Phi is an important movement that leads to novel philosophically significant results and it is amazing how philosophy and experimental psychology can work together. We shall see, however, that the picture of X-Phi leading to a revolution within philosophy that leaves behind burnt armchairs and outdated theories is highly mis-

Philipp Berghofer, Karl-Franzens-Universität Graz
philipp.berghofer@uni-graz.at

https://doi.org/10.1515/9783110664850-005

leading. X-Phi is neither revolutionary in the sense of making traditional philosophical methodology obsolete, nor is it novel in the sense of offering a hitherto never conducted way of doing philosophy.[1] In this section, we shall see how X-Phi can supplement traditional philosophical research and could particularly lead to progress in phenomenological research. In the following sections, we show that the methodology of X-Phi is no invention of the 21st century, but that it has been practiced or at least anticipated in Meinong's Graz school of experimental psychology, exemplified particularly in the works of Vittorio Benussi.

So what is X-Phi? Although there is no universally accepted answer to this question, the term as such suggests the following definition:

DX-Phi: Conducting experiments with the aim of making genuinely philosophical contributions.

DX-Phi accounts for the fact that X-Phi is more specific than an empirically informed philosophy. *Experimental* philosophers do not only use empirical data, they *conduct their own empirical research* to gain empirical data. Also, experimental *philosophers* do not just gain experimental data that has philosophical implications, they gain data *in order to* draw philosophical conclusions. Arguably, any work in science has philosophical implications. The work of physicists doing empirical research that leads to new insights about the nature of space, time, and matter has important philosophical implications, but we do not want to classify genuinely physical research as a version of X-Phi.

DX-Phi captures the basic idea of X-Phi and since it does not specify what kind of experiments experimental philosophers are doing to gain philosophically significant results, it is broad enough to cover all philosophically motivated experimental research. However, one might think that DX-Phi is too broad since experimental philosophers do not do the kind of experimental research we know from physics and chemistry. In fact, it is one specific type of experimental meth-

1 By traditional philosophical methodology, I mean philosophy as an a priori "armchair" discipline that uses thought experiments and gives epistemic weight to philosophical intuitions such as Gettier intuitions. With respect to X-Phi, we may distinguish between a strong and a modest approach. According to the strong approach, traditional philosophical methodology is obsolete and needs to be *replaced* by purely empirical methods such that all of philosophy becomes some sort of X-Phi. According to the modest approach, traditional philosophical methodology can be *supplemented* by experimental investigations such that both methodologies complement each other. The present paper argues that the works of Meinong and especially Benussi exemplify the modest approach. The strong approach, in my opinion, is a dead end.

odology that virtually every experimental philosopher is using: the method of experimental psychology. Accordingly, the basic idea of X-Phi has been expressed as follows:

> Experimental philosophy is the name for a recent movement whose participants use the methods of experimental psychology to probe the way people make judgments that bear on debates in philosophy.[2]

Thus, we may define X-Phi as follows:

NDX-Phi: Engaging in experimental psychology in order to gain philosophically significant results.

NDX-Phi captures the basic idea of X-Phi and specifies the method that is actually used by experimental philosophers. However, one might argue that NDX-Phi is still too broad since it does not specify the kind of object that is experimentally investigated. Again, most of current X-Phi has a very specific object of investigation: intuitions.[3] Many arguments in philosophical debates have the following structure: P is intuitively plausible, thus we have a prima facie reason for accepting p.

However, a philosopher's intuition most often is the intuition of a white male with a Western cultural background and philosophical background assumptions. 21st century X-Phi was launched with the aim of empirically investigating whether philosophers' intuitions are shared by "normal" people from various backgrounds.[4] While, of course, not all of X-Phi is of such a form, it is still true, after nearly two decades of research, that the primary objects of investigation in X-Phi are people's intuitions. Accordingly, X-Phi is often specified as follows:

> Since its inception, experimental philosophy has been bound up with methodological questions regarding the status of intuition. [...] Some experimental philosophers conceive of

---

2 Nadelhoffer / Nahmias 2007, p. 123.

3 By intuitions experimental philosophers usually understand something like gut feelings or immediate judgments. The idea is to make surveys and to test people's reactions to hypothetical scenarios. For instance, people are confronted with a description of a Gettier case and asked whether the person in question knows that *p* or only believes that *p*.

4 As a side note, I would like to point out that this experimental method has already been practiced by Jean Buridan in the 14th century, when he challenged the claim of Nicholas of Autrecourt that no one could doubt the law of non-contradiction by actually testing the intuitions of non-philosophers. See Maier 1967, p. 392.

> their project as involving systematic empirical study of intuitions qua intuitions, of the psychological processes that produce them, and of the conceptual frameworks that they reflect.[5]

Hence, a more narrow definition of X-Phi:

MNDX-Phi: Engaging in experimental psychology in order to investigate people's intuitions and draw philosophically significant conclusions.

MNDX-Phi pretty much covers what experimental philosophers are doing in the 21st century. They empirically investigate "our" intuitions and from their empirical investigations they draw philosophical implications. One can distinguish different movements within X-Phi based on what kind of implications they draw. In this context Nadelhoffer and Nahmias make the following classification: Results of X-Phi have been used (i) to determine *what* people think in order to reinforce philosophical theories, (ii) to determine *how* people think in order to shed light on psychological mechanisms producing people's intuitions, and (iii) to challenge the reliability of philosophical intuitions by showing that they depend on epistemically irrelevant factors.[6] Especially with respect to (ii), the lines between X-Phi and the cognitive sciences become blurred, but this is not our worry here. I briefly want to address (iii), which is known as the negative program of X-Phi.

The negative program is the best-known but also most controversial branch of X-Phi. The success of the negative program, of course, would have huge philosophical implications and could be seen as a revolution of philosophical methodology. This is because regarding orthodox philosophical methodology, there is considerable agreement that "intuitions are presented as our evidence in philosophy"[7] and that "analytic philosophy without intuitions just wouldn't be analytic philosophy".[8] The reliance on intuition is considered as one of the defining features of philosophy:

> One thing that distinguishes philosophical methodology from the methodology of the sciences is its extensive and avowed reliance on intuition.[9]

5 Nado 2014, p. 11.
6 See Nadelhoffer/Nahmias 2007, p. 126.
7 Williamson 2007, p. 214.
8 Weinberg 2007, p. 318. See also Pust 2000, p. xiii.
9 Goldman 2007, p. 1.

If the negative program were successful in showing that our reliance on intuitions is unjustified, this would shatter our current understanding of how (analytic) philosophy works.[10]

However, there are three things to note. First, the negative program might be the best-known branch of X-Phi, but there is *no systematic reason* why a proponent of X-Phi should subscribe to the negative program. Rejecting the negative program is compatible with being an experimental philosopher. Secondly, as a matter of fact, most experimental philosophers do *not* engage in the negative program. The negative program "has attracted attention disproportionate to its size"[11] and Joshua Knobe has pointed out that it only "accounts for 1.3% of the studies reported over the past five years".[12] Thirdly, some experimental philosophers have recently used experimental methods to argue *against* the negative program, either by rerunning experiments that were deemed to imply the unreliability of intuitions or by conducting new experiments that suggest that (some of) our intuitions are reliable after all.[13] Thus, we can introduce a fourth movement within X-Phi, namely (iv) to argue against the negative program of X-Phi.

This last point highlights a problem that could be seen as the major shortcoming of X-Phi. The problem is that in many cases a philosophical theory t is supposed to be supported by empirical results, there seem to be opposing empirical results that suggest ~t. These recent developments cast serious doubt on the viability of X-Phi. Let me illustrate this problem by discussing the most popular work of the negative program.

Weinberg et al. (2001) is widely considered the birth of (modern) X-Phi. It is also the birth of the negative program. Their survey indicated that while students with a Western cultural background share the intuition that Gettier cases are not cases of knowledge, students with an East Asian cultural background do *not* share this intuition (to the same extent).[14] If it was true that Gettier intuitions are influenced by an epistemically insignificant factor such as cultural background, this would undermine arguments based on such intuitions. Since in modern de-

---

10 To be sure, there are scholars such as Cappelen 2012, Deutsch 2010, and Ichikawa 2014 who actively and explicitly dispute the claim that intuitions are essential for philosophical inquiry. However, these authors are clearly in the minority.

11 Williamson 2016, p. 22.

12 Knobe 2016, p. 39.

13 Cf. Boyd / Nagel 2014 and Wright 2010.

14 As first pointed out by Roderick Chisholm, it is interesting to note that already in Meinong one can find something close to a Gettier case. See Meinong 1906, p. 398f.

bates Gettier intuitions are considered the prime examples of justification-conferring philosophical intuitions, this attack of Weinberg et al. has been taken extremely seriously.

The problem now is that the empirical results of Weinberg et al. (2001) have in turn been challenged *on empirical grounds.*

> To date, there is no robust evidence that the epistemic intuitions of different demographic groups are deeply at odds with each other. Although the Weinberg et al. (2001) results have been heavily cited, it is doubtful that they are robust. [...] An effort to replicate the Weinberg et al. (2001) Gettier case results with multicultural North-American participants failed to detect ethnically correlated differences in responding to Gettier cases.[15]

Other examples of such conflicting empirical results are works on the correlation between CRT and the Knobe Effect[16] and the question of whether people's intuitions support compatibilism or incompatibilism.[17]

The first attack on X-Phi was launched by Kauppinen.[18] Kauppinen argues that X-Phi "rests on mistaken assumptions about the relation between people's concepts and their linguistic behaviour".[19] He holds that the method of X-Phi, namely surveying people's intuitions, fails *in principle* since, among other things, the Socratic way "of engaging in dialogue with the test subjects" is necessary to avoid performance errors but "violates the spirit and letter of experimentalist quasi observation".[20] It would go beyond the scope of this paper to assess Kauppinen's criticism in detail. It ought to be pointed out, however, that even if all Kauppinen says was true, this would not undermine X-Phi as such but only the kind of X-Phi that investigates people's philosophical intuitions by doing surveys. A similar objection against X-Phi has been put forward by Simon Cullen, who argues that X-Phi's underlying "assumption that intuitions can be simply read off from survey responses" is not accurate.[21] Cullen calls this widespread method in X-Phi to survey people's intuitions and draw philosophical implications from the results "Survey-Driven Romanticism".

---

15 Boyd / Nagel 2014, p. 115. Cf. also Seyedsayamdost 2012, Turri 2013, and especially Machery et al. 2017.

16 Cf. Cokely / Feltz 2009 and Pinillos et al. 2011.

17 See Nichols / Knobe 2007.

18 Kauppinen 2007.

19 Kauppinen 2007, p. 95.

20 Kauppinen 2007, p. 106.

21 Cullen 2010, p. 275.

> According to Survey-Driven Romanticism, people's philosophical intuitions are implanted within them in some way, and by administering simple surveys we can discover them.[22]

For Cullen, survey responses must not be identified with intuitions and doing so means "ignoring the established social and cognitive science literature on survey methodology".[23]

Again, it is to be noted – and Cullen is very well aware of this – that even if Survey-Driven Romanticism fails, this does not undermine all of X-Phi. Most *but not all* of X-Phi is concerned with drawing conclusions from people's intuitions. (This is why MNDX-Phi is *too* narrow.) There simply is no systematic reason why X-Phi should be obliged to focus on intuitions.[24] In fact, the reasons why much of X-Phi has focused on making surveys and aimed at exposing people's intuitions are purely contingent: It is easy and it is cheap.[25] I shall conclude this section by briefly pointing out in which areas X-Phi should go beyond survey-driven examination of intuitions and how such a future X-Phi could supplement phenomenological research.

Many arguments in philosophical debates have the form: Intuitively, we would all agree that p, thus we should accept p. X-Phi can make important contributions in making such claims more honest by testing them empirically. Furthermore, and perhaps more importantly, X-Phi can show which of our intuitions are more robust and help to discern factors that influence our judgments negatively.[26] While early works in current X-Phi suggested that experimental investigations reveal that orthodox philosophical methodology is inherently mistaken, it seems that it has become the predominant view among experimental philosophers that "experimental methods can complement and even strengthen armchair-style philosophy".[27] However, there are crucial philosophical problems that cannot be tackled by surveying people's intuitions. For instance, problems where we have no reason to believe that people have reliable intuitions about. It would be strange to discuss the implications of Gödel's incompleteness theorems on the philosophy of mathematics by reviewing people's intuitions. Something similar is true for the most common phenomena. Everybody knows what it is like to have

---

22 Cullen 2010, p. 277.

23 Cullen 2010, p. 278.

24 In section 3, we shall see that the works of Benussi exemplify X-Phi that is not limited to the study of intuitions.

25 See Prinz 2008, p. 200.

26 Following Cullen, however, we have to keep in mind that such undertakings are not as straightforward as it may seem.

27 Mortensen / Nagel 2016, p. 53.

a perceptual experience. However, the nature of perception and the questions of what it is that distinguishes perceptual experiences from other experiences and what it is that makes experiences justifiers are questions where surveying people's intuitions is only of limited use. In this context Metzinger and Windt claim that an intuition-based X-Phi cannot tell us anything about the nature of phenomenal consciousness or intentionality.[28] Thus, Metzinger and Windt argue for what they call an intuition-free X-Phi.

According to them, an intuition-free X-Phi could be of particular signifycance in the field of philosophy of mind. While here the boundaries between the cognitive sciences and philosophy are especially blurry, what makes such an intuition-free X-Phi a (partly) philosophical undertaking is the aim of conducting experiments in order to tackle philosophical questions.[29] An intuition-free X-Phi is any X-Phi that does not merely consist in surveying people's intuitions. Usually, intuition-free X-Phi is experimentally more sophisticated. The aim is to learn something about the human mind that is of philosophical significance.[30] In section 3, we shall see that Vittorio Benussi exemplifies this kind of X-Phi and that his results are not only important for philosophy of mind but also for epistemology.

I believe that especially phenomenological investigations could benefit from an intuition-free X-Phi as such a version of X-Phi seems to be well-suited to experimentally investigate the human mind, including basic types of experiences like perceptual experiences.[31] Such experimental research may help to tackle the following questions: How does perceptual learning shape the way we perceive?[32] What do cases of perceptual learning tell us about whether perceptual experiences can be a source of immediate justification even if they depend for their justificatory force on factors such as the existence of foregoing experiences?[33] Does perception necessarily include judgment?

Contributing to answering such questions would undoubtedly be of philosophical significance. Let us summarize the main points of this section.

---

28 Metzinger / Windt 2014, p. 282.

29 See Metzinger / Windt 2014, p. 311.

30 For a recent example of an intuition-free X-Phi see Lenggenhager et al. 2007.

31 To be sure, in no way do I want to suggest that experimental research could *replace* Husserlian-style phenomenology. For a refutation of such an undertaking cf. Rinofner-Kreidl 2004. Furthermore, I do not want to suggest that Husserlian-style phenomenology can be naturalized. All I argue is that X-Phi (and especially an intuition-free X-Phi) can supplement phenomenological research.

32 For a related question, namely how attention can alter perceptions, see Tse et al. 2013.

33 See Berghofer 2018, p. 21.

1) Broadly speaking, doing X-Phi can be defined as conducting experiments with the aim of making genuinely philosophical contributions (DX-Phi).
2) X-Phi should not be understood as a rival to orthodox philosophical methodology but as a methodological complement.
3) Although most of current X-Phi consists of survey-based investigations of people's intuitions, there is no systematic reason why X-Phi should be limited to such a methodological approach.
4) An intuition-free X-Phi could especially contribute to phenomenological investigations by investigating perception and intuition.

2 Meinong and Experimental Philosophy

It is safe to say that Meinong is best known for two things: His object theory and establishing the first laboratory for experimental psychology in Austria. Meinong's object theory is his version of an intentionalist philosophy that distinguishes his philosophical system from the immanence doctrine of his teacher Franz Brentano and from the phenomenological approach of Edmund Husserl.[34] The Graz laboratory for experimental psychology may be best known for experimentally reinforcing the so-called theory of production. While Stephan Witasek and Rudolf Ameseder were primarily working on the theoretical foundations of the theory of production, it was most notably Vittorio Benussi who delivered experimental data in support of the theory.

Since Meinong is the founder of the Graz school of object theory and experimental psychology, the question arises whether Meinong can be viewed as an experimental philosopher. Did Meinong use methods of experimental psychology in order to draw philosophical conclusions? Or are philosophy and experimental psychology for Meinong two totally distinct disciplines that lack significant concatenations? The second question can easily be denied. As we shall see, for Meinong "experimental psychology and philosophy were areas of research intrinsically interconnected".[35] This makes Meinong a truly interdisciplinary researcher who aims at fruitfully bringing together philosophy and experimental

34 Although there are important differences between Meinong and Husserl, their philosophical systems also share crucial characteristics such as being a philosophy from below. "Although Meinong's theory of objects is an a priori science, because it has to do with whatever can be known a priori about objects, its method can be characterized as a *philosophy from below*, a working method that follows the precepts of a kind of bottom-up thinking" (Marek 2013). See also Rinofner-Kreidl 2000, p. 256, 311–344.

35 Albertazzi 2013, p. 580.

psychology and may distinguish him from other early experimental psychologists.

In this section, however, we shall see that Meinong cannot be straightforwardly regarded as an experimental philosopher in line with definition DX-Phi brought forward in the previous section. This is because in his research Meinong used philosophy and experimental research mostly one-way: Object theory was used as the theoretical foundation of experimental research. There is no clear example in Meinong's oeuvre where he uses experimental results of his Graz school explicitly in order to argue for a philosophical theory.[36] However, it is safe to say that Meinong paved the way for experimental philosophy. As we shall see in the next section, this is because Meinong's pupil Vittorio Benussi can be viewed as an experimental philosopher, perhaps the first one in the 20th century.[37]

The first thing to note is that there is no doubt that for Meinong experiments can make crucial contributions to philosophical investigations. In his *Über die Erfahrungsgrundlagen unseres Wissens* Meinong repeatedly refers to Locke's famous description of an experiment in which one hand is put in warm water while the other is put in cold water. Then, both hands are placed in a bottle of lukewarm water. As a result, the same water feels warm to one but cold to the other hand. For Meinong, this shows that we have to distinguish between sensation and perception[38] and that judgments of comparison (*Vergleichungsurteile*) cannot possess evidence for certainty (*Gewissheitsevidenz*) but only evidence for presumption (*Vermutungsevidenz*).[39]

---

36 However, it should be pointed out that Meinong refers to experimental data of the Graz laboratory when he discusses the relationship between objects of higher order and inner perception (Meinong 1899, p. 437ff.). I am grateful to Venanzio Raspa for pointing me towards this passage.

37 Stich and Tobia point out that "[p]hilosophically motivated ethnography like Richard Brandt's pioneering study of Hopi ethics (Brandt 1954) and John Ladd's study of the moral code of the Navaho (Ladd 1957)" qualifies as X-Phi in a broad sense, and they argue that "Brandt and Ladd have a good claim to being the first important contributors to contemporary experimental philosophy" (Stich / Tobia 2016, p. 5). I take it that the present paper shows that the early experimental psychologists such as Meinong and Benussi, who clearly predate Brandt and Ladd and are methodologically closer to current X-Phi, have a more rightful claim to this title.

38 See Meinong 1906, p. 411.

39 See Meinong 1906, p. 462. Note that Meinong allowing for the possibility of Vermutungsevidenz shows that his conception of evidence (and immediate justification) is not as strict as Brentano's. Here Meinong is in agreement with Husserl since both allow that evidence is also possible with respect to physical objects. While Husserl defines his notion of *inadequate evidence*

Although Meinong notes that Locke's experiment is "primitive" compared to "our psychological laboratories" and does not look a lot like "future experimental psychology", he leaves no doubt that it is of philosophical significance.[40] This shows that Meinong clearly *advocates* and *uses* X-Phi. He holds that experimental psychology can contribute to philosophical discussions and he uses X-Phi to draw philosophical conclusions. Of course, this does not show that Meinong *engages* in X-Phi. Meinong is a philosopher and an experimental psychologist who believes that both disciplines are interconnected and that philosophy can benefit from experimental data. But are there clear examples in which Meinong sets up an experiment to draw philosophical conclusions?

In this context it is important to point out that Meinong clearly allows that experimental data can supplement or refute theoretical results. This can be seen when Meinong in his "Über die Bedeutung des Weberschen Gesetzes" draws conclusions about the measurability of the mental based on his investigations on Weber's law.[41] This work has been very positively reviewed by Russell 1899.[42] However, it is not straightforward that these conclusions are genuinely philosophical. More importantly, in this case, Meinong does not base his conclusions on experiments conducted in his laboratory.

So what we are looking for is a clear example in which Meinong uses experimental results of the Graz school to draw genuinely philosophical conclusions. Importantly, combining experimental-psychological and philosophical research is a distinctive feature of Meinong's Graz school.

> Therefore the "Experimental-Psychologische Apparat", especially when suitably combined with a Philosophical Seminar, would be an indispensable basis for advanced discussions or related exercises in a fundamental field of philosophical sciences.[43]

Albertazzi provides the following picture:

---

phenomenologically (i.e., what it is like to undergo, for instance, a perceptual experience in which the physical object is only given inadequately, namely in adumbrations), Meinong, in his late but extensive *Über Möglichkeit und Wahrscheinlichkeit*, links his conception of Vermutungsevidenz to a theory of probability (cf. Höfer 2014).

40 Meinong 1906, p. 411.

41 See Haller 1995/96, p. 33.

42 For how Meinong's investigations on Weber's law have been reviewed by Russell and had an impact on Russell's work, see Guigon 2005, section 2.

43 Meinong to the Austrian Ministry (draft) in 1893; cited in Dölling 2001, p. 61.

> The complementarity between *theory* and *experimentation* was therefore a characteristic feature of the theory of Meinong's school; [...] As Meinong writes in the preface to *Untersuchungen zur* Gegenstandstheorie *und Psychologie*, in 1904, the book which presented the philosophical and psychological researches of his pupils, and celebrated the tenth anniversary of the opening of his laboratory of psychology, "[w]e have always been of the opinion that experimentation in psychology is never an end in itself".[44]

When Meinong says that "experimentation in psychology is never an end in itself" this suggests that he performs experimental psychology to draw theoretical and ultimately philosophical conclusions.[45] Especially with respect to his object theory, Meinong clearly holds that there are significant links to experimental psychology: "And thus object theory has been no stranger to the Grazer psychological laboratory."[46] Does this mean that under the direction of Meinong the Graz school conducts experiments whose data are used in theoretical writings to back up object theory? This does not seem to be the case. When Meinong talks about the relationship between object theory and experimental psychology, the kind of impact he has in mind is mostly one-way: Object theory is considered the theoretical foundation and framework of experimental research. Accordingly, when Meinong points out that object theory has been no stranger to the Grazer experimental laboratory this does not mean that experiments were conducted to back up object theory but that insights of object theory were used to support experimental research. In this sense Meinong says "that object-theoretical knowledge and skills can very well be useful to psychological research".[47] Antonelli and Manotta have pointed out that at the Graz school psychological issues were treated within the framework and assumptions of object theory.[48] Albertazzi has argued that such a philosophical framework was a burden for psychological research and was "one of the reasons for the oblivion into which the Graz psychologists sank".[49]

Thus, while Meinong assumes that experimental psychology and philosophy can profit from each other, the way in which object theory and experimental psychology have been linked is largely one-sided: Insights of object theory are used to advance research in experimental psychology, but it does not seem to be the case that Meinong runs experiments to boost object theory. This does not

44 Albertazzi 1995, p. 322.
45 Meinong 1904a, p. VII.
46 Meinong 1904a, p. VII.
47 Meinong 1904b, p. 46.
48 See Antonelli / Manotta 2009, p. 126f.
49 Albertazzi 2001, p. 321.

imply that Meinong holds that object theory cannot in principle profit from running experiments but simply that Meinong has not exemplified how this could be done.

Let us summarize the findings of this section: Following Brentano, Meinong adopted an empiricist approach towards philosophy that methodologically focuses on introspection within the framework of descriptive psychology. A distinctive feature of a Brentanian philosophy is that the intentionality of consciousness is considered the starting point of all philosophical investigations. One distinguishing factor between Meinong and Brentano is that Meinong supplements descriptive psychology with experimental psychology.[50] Meinong, in contrast to Brentano, is actually involved in experimental research.[51] Furthermore, Meinong holds that experimental psychology is not an end in itself but can be used to complement and correct theoretical investigation, for instance, results within descriptive psychology. At first glance, one might think that Meinong obviously has to be considered an experimental philosopher. He is a philosopher who conducts experiments. The question is, however, whether he draws *philosophical implications from his experiments.*

In the literature, it has been pointed out that Meinong has at least occasionally used experimental results in order to make or correct theoretical claims as for instance when he draws conclusions about the (measurability of the) mental based on his investigations on Weber's law. Such findings are important in the field of psychophysics and while it may be true that there are significant philosophical implications, this is not enough to classify Meinong as an experimental philosopher. This is mainly for one reason: There is no clear indication that Meinong's results are based on experiments that he has conducted or supervised himself.

Since Meinong has insisted that there are significant relations between object theory and experimental psychology, we have investigated whether Meinong has used empirical data to back up the centerpiece of his philosophical system. It has turned out, however, that for Meinong object theory has to guide experiential investigations. There is no clear example of Meinong explicitly referring to experiments *in order to back up object theory*. Of course, one might argue that since Meinong sets up experiments according to insights of object theory, success of

50 See Antonelli / Manotta 2009, p. 125f.

51 As a side note, it should be mentioned that Brentano was interested in doing experimental research. In 1874, he unsuccessfully requested a psychological laboratory in Vienna. Had his request been successful, this laboratory would have even predated Wundt's. See Münch 1997, p. 104.

his experiments may be viewed as an indirect confirmation of object theory, or that results of Meinong's earlier experiments had some influence in leading him towards object theory. Again, however, this seems to be too little to view him as a clear example of an experimental philosopher. Meinong may be a precursor of X-Phi, but there is no clear evidence that we should regard him as having practiced X-Phi.

3 Benussi and Experimental Philosophy

In the previous section, we have seen that, while Meinong believes that philosophy and experimental psychology can supplement each other, he cannot be straightforwardly regarded as an experimental philosopher since there are no clear cases where he draws philosophical conclusions from experiments conducted at the Graz school. In this section, we shall see that one of Meinong's most gifted students, Vittorio Benussi, better qualifies as an experimental philosopher. There is an interesting asymmetry between Benussi and Meinong. Concerning the relationship between philosophy and experimental psychology, Meinong is, so to speak, more philosopher than experimental psychologist. From the beginning of the 20th century onward, he is primarily working on philosophical issues, and his object theory is not only the centerpiece of his philosophical system but also provides the theoretical framework for his experimental research.[52] Benussi, on the other hand, is clearly more experimental psychologist than philosopher. His work and life are dedicated to experimental research, and while he is concerned with theoretical and philosophical questions, he is at pains not to let theoretical prejudice spoil experimental research. In this context Dölling claims: "Benussi, by contrast, was more a specialist in scientific psychology, who had little interest in philosophy".[53]

We will qualify this statement that Benussi only "had little interest in philosophy" in a moment. What is true is that Benussi was *the* experimental psychologist of the Graz school. This is evident, for instance, with respect to the Graz theory of production. While Meinong, Witasek, and Ameseder were more concerned with theoretical or descriptive analyses, Benussi was the one who was gaining

52 It should be noted, however, that for Meinong philosophy is not a "sharply delineated science but an assemblage of sciences which are held together by the fact that they have to do with mental phenomena. [...] Because of this, psychology becomes a fundamental discipline of this philosophical congeries" (Marek 2013).
53 Dölling 2001, p. 62.

experimental data in support of the theory.[54] Benussi's modus operandi is distinguished by his "theoretical minimalism"[55] and shaped by the intention "to replace the evidence of thought with the evidence of facts".[56] Benussi has been called "the most productive and effective experimental psychologist that Austria had had"[57] and his experimental skills have been compared to Leonardo's.[58]

However, from the facts that Benussi was first and foremost an experimental psychologist and that he kept philosophical assumptions to a minimum when conducting his experiments, we should not conclude that he had no interest in philosophy or that he did not draw philosophical conclusions from his experimental works. In his *Autobiografia scientifica* Benussi states that

> even in the most detailed investigations, despite all experimental labyrinths, the broadest purely theoretical-philosophical problems have been kept in sight.[59]

This shows that Benussi was interested in philosophical problems and that his experimental works cannot be totally detached from philosophical issues. It is true, however, that Benussi's focus clearly is on the experimental work and *not* on discussing philosophical implications of his experiments. One can only speculate why this is so. Maybe the reason really is a lack of interest or the fact that some of his main results are incompatible with core theoretical beliefs of his teacher Meinong or a combination of both.

In this context we shall see that there is no doubt that Benussi has been aware that some of the theoretical-philosophical conclusions he draws from his experiments are diametrically opposed to central teachings of Meinong but bring him remarkably close to Husserl's phenomenology. There are four theoretical-philosophical conclusions in particular that I take to be of significance.

P1: Based on his experimental works on inner-time consciousness,[60] Benussi argues for "an inextricable correlation of time and consciousness, active subjectivity, and the constitution of the temporal horizon".[61]

---

54 See, e.g., Benussi 1904, § 23.
55 Antonelli 1994, p. 87.
56 Fondo Benussi Carteggio, 9, autobiografia scientifica, archivio storico della psicologia italiana (University of Milan), cited in Antonelli 1994, p. 14.
57 Boring 1929, p. 439.
58 Antonelli 1994, p. 28.
59 Fondo Benussi Carteggio, 9, autobiografia scientifica, archivio storico della psicologia italiana (University of Milan), cited in Antonelli 1994, p. 3.
60 Benussi 1913.
61 Antonelli 1994, p. 113.

P2: Through his experimental investigations of optical illusions such as the Müller-Lyer illusion and reversible images (Vexierbilder), Benussi developed an understanding of formal plurivocity (Gestaltmehrdeutigkeit) that led to a conception of intentionality according to which consciousness plays a productive and constituting role.[62]

P3: In his laboratory, Benussi managed to induce brute emotional states that he considered to be emotional acts that are *not* founded on cognitive acts.[63]

P4: Based on his empirical studies of optical illusions, Benussi concludes that perception does not involve judgment.[64]

P1 and P2 exemplify how Benussi's experimental investigations bring him closer to central points of Husserl's phenomenology, P3 and P4 how he turns away from central teachings of Meinong. P1–P4 refer to philosophically significant conclusions Benussi draws from his experimental studies. Thus, Benussi seems to qualify as an experimental philosopher. One might doubt that P1–P4 are really philosophically significant. Should P1–P4 not be considered psychological rather than philosophical findings? However, similar objections could be raised to current X-Phi. The Knobe effect, for instance, is one of the prime examples of research in X-Phi, but is it a straightforward philosophical result? In X-Phi the boundaries between philosophical and psychological investigations become blurred. This is true for the works of Benussi as much as for the works of current experimental philosophers.

Finally, we shall briefly discuss P4. By doing so, we will see how Benussi's experimental research leads him to the conclusion that the Brentano-Meinong thesis that perception essentially involves judgment is mistaken and how Benussi's findings are relevant to current debates in epistemology.

For Meinong, it is clear that "all perceptions [...] are judgments of existence [Existenzurteile]".[65] It is part of the essence of perception to be a judgment of existence such that the content of the judgment corresponds to the content of the

62 See, e.g., Benussi 1904, p. 414 and Antonelli 1994, p. 114–120. Antonelli summarizes: "Benussi, der sich hierbei nicht zufällig immer mehr Husserls Phänomenologie annähert, entwickelt den Brentano-Meinongschen Intentionalitätsbegriff von einem rein klassifikatorischen Kriterium zu einem leistungsfähigen Prinzip des Bewußtseins, dessen Charakter sich als produktiv erweist" (Antonelli 1994, p. 118).

63 See Vendrell Ferran 2009, p. 232.

64 See Benussi 1904.

65 Meinong 1906, p. 388.

act of perception.[66] When you perceive a table in front of you, this act of perception involves the judgment "There is a table". This doctrine that perception essentially involves judging frequently occurs in central passages in Meinong's works and also plays an important role for Brentano. Based on his experimental studies, Benussi concludes that this view is mistaken. Most of Benussi's works at his time in Graz focused on optical illusions. In his dissertation, he experimentally studied the Zöllner illusion and three years later he published an article entitled "Zur Psychologie des Gestalterfassens (Die Müller-Lyersche Figur)". This article was published in 1904 in Meinong's famous anthology *Untersuchungen zur Gegenstandstheorie und Psychologie. Zum zehnjährigen Bestande des Psychologischen Laboratoriums der Universität Graz* which received considerable international attention. Here and in many of his later works, Benussi argues that the received view on optical illusions is fundamentally mistaken. The German term for optical illusions is "optische Täuschungen". A "Täuschung" is a delusion. However, Benussi insisted that illusions such as the Müller-Lyer illusion are *no delusions at all.* A delusion is an error and an error can only occur with respect to a judgment.[67] An optical illusion, according to Benussi, should not be viewed as an error of judgment, which is why optical "delusions" are no delusions. Benussi prefers the term "inadequate representation".[68]

The point is that Benussi's detailed studies on how people react to optical illusions and how such effects can be increased and reduced led him to the conclusion that the essence or nature of perception does not involve judgment.[69] About twenty years later, he complemented his analysis of perception by arguing that the nature of perception is best disclosed by the term "presence". "To perceive an object means to experience its presence".[70] Perception, by its very nature, has the character of "presence". Most importantly, perceptual experiences have this character of presence even if the intended object does not exist, "is 'physically' absent".[71] This means that the character of perception is determined phenomenologically. Perception has the character of presence because, in contrast to other types of experiences such as believing, hoping, imagining, etc., in

66 Meinong 1888, p. 118.

67 Benussi 1904, p. 391.

68 Benussi 1904, p. 391. See also Benussi 1906.

69 See Benussi 1904, p. 391.

70 "Percepire un oggetto è viverne la presenza" (*La suggestione e l'ipnosi come mezzi di analisi psichica reale*, cited in Antonelli 1994, p. 56, n. 59). See also Antonelli 1994, p. 163.

71 Antonelli / Manotta 2009, p. 149. With respect to "known illusions" such as the Müller-Lyer illusion, this means "that the character of presence does not change even if the observer *knows* that" her perceptual experience is illusionary (Antonelli 1994, p. 163).

perception it seems to you that the object of your experience is present to you. "[T]he object is immediately [unmittelbar] and vividly [anschaulich] 'there,' 'in front of us'".[72] To put it differently, *from the first person perspective* perception distinguishes itself by making its object present to you. In current terminology, we may say that perception has a distinctive *phenomenal character*,[73] namely, according to Benussi, the *character of presence*.

This, of course, demonstrates again that Benussi's experimental investigations have brought him considerably close to Husserl's phenomenology. For Husserl, perceptions are originary presentive intuitions, which means that they present their objects as "bodily present"[74] and give them in a "fleshed out" manner.[75] Furthermore, Husserl, just like Benussi, points out that not only veridical perceptions have the character of originary givenness (Husserl) or presence (Benussi) but also hallucinations and illusions.[76]

It is to be noted that Benussi is aware of the fact that his understanding of perception differs from Meinong's[77] and that there is a remarkable affinity to Husserl.[78] Let me briefly highlight why this experimental finding of Benussi that the nature of perception does not involve judgment but is disclosed by its distinctive phenomenal character of presence is of great epistemological signifycance.

In current debates it is often argued that perceptual experiences are a source of justification but are not in need of justification themselves.[79] Seeing a black laptop in front of me justifies me in believing that there is a black laptop. The act of seeing justifies but is not in need of being justified. Thus, immediate justification is possible and the threat of an indefinite regress is avoided. However, if perception is itself a judgment, it seems that perception is in need of justification. If perceiving a laptop means judging that there is a laptop, the question arises, what is the justifier for this judgment. Thus, Benussi can but Meinong cannot allow for the very plausible idea that perception provides immediate justification.

72 Antonelli 1994, p. 163.
73 "It is definitional of experience, as the term is used here, that they have some phenomenal character, or more briefly, some phenomenology. The phenomenology of an experience is what it is like for the subject to have it" (Siegel 2016).
74 Husserl 1973, p. 14.
75 Husserl 1984, p. 458.
76 See Husserl 1973, p. 15. See also Erhard 2012, p. 57 who refers to this passage.
77 Cf. an unpublished note of Benussi quoted in Antonelli 1994, p. 57f.
78 Cf. a note of Benussi translated into German in Antonelli 1994, p. 58.
79 See Huemer 2001, p. 97 and Ghijsen 2016, p. 37f.

Interestingly enough, one of the main arguments in current debates for why perception (or what has been termed perceptual seemings) cannot be reduced to beliefs is motivated by the Müller-Lyer illusion.

> For instance, when someone who is familiar with the Müller-Lyer illusion is looking at the Müller-Lyer lines, he will have the perceptual experience that the lines are unequal, but he will not believe that the lines are unequal. This provides some reason to think that perceptual experiences are distinct from beliefs.[80]

For philosophers who hold that perceptual experiences are a source of immediate justification, it is central to point out that they are distinct from beliefs, and the Müller-Lyer illusion is considered the prime example for demonstrating this fact. Importantly, philosophers usually do not conduct experiments to back up this claim with experimental data. Nor do they refer to experimental studies. We may thus say that Benussi's investigations can provide *experimental support* for such claims.

Finally, I would like to point out that there are current epistemologists who argue (i) that perceptual experiences have the character of "presentationality"[81] or a "presentational phenomenology"[82] and (ii) that perceptual experiences gain their justificatory force simply by virtue of this distinctive phenomenal character. While Benussi is silent with respect to the second claim, his analysis of perception as having the character of presence perfectly resembles the claim that perceptual experiences have a distinctive phenomenal character which can be cashed out by terms such as "presentational".

Interestingly, Benussi's experimental research qualifies as the kind of X-Phi we have labeled intuition-free X-Phi in section 1. Benussi does not survey people's intuitions or gut reactions about hypothetical scenarios but aims at a more direct experimental investigation of perception that includes cases of perceptual learning and can be used to draw conclusions about the nature of perception. (According to Benussi, the nature of perception is disclosed by the character of presence.) Benussi's kind of experimental research has been termed a genetic-experimental phenomenology (genetische Experimentalphänomenologie[83]) and indeed we have seen that his experimental research can be regarded as an experimental confirmation of some of Husserl's central theses.[84]

---

80 Ghijsen 2016, p. 37.

81 Bengson 2015.

82 Chudnoff 2013.

83 Antonelli / Manotta 2009.

84 It is to be noted that Benussi is not the only early experimental psychologist with a close affinity to some of Husserl's main theses. This seems also to be true for the Würzburg school and

4 Conclusion

Experimental philosophy is a novel branch within philosophy that allows gaining philosophically important results by experimental methods. In this way, X-Phi can fruitfully supplement traditional philosophical methodology. However, in section 1 we have seen that most of current X-Phi consists of surveying people's intuitions and that there are reasonable reservations about this method. We have argued that current X-Phi could benefit from bringing an intuition-free X-Phi into focus that not via surveys but more directly investigates different types of experiences. In section 2, we have posed the question of whether Meinong can be considered an experimental philosopher. Meinong is a philosopher and experimental psychologist who deems both disciplines to be internally related such that both disciplines can benefit from each other. However, Meinong does not use experimental data acquired by himself or by his Graz psychological laboratory to back up his philosophical views. It is the other way around. He uses insights from his object theory to support experimental investigations. Thus, Meinong is more of a precursor than a practitioner of X-Phi. Meinong's pupil Benussi, however, clearly draws philosophically important conclusions from his experimental investigations. Although Benussi is definitely more psychologist than philosopher, he uses his empirical data to argue for philosophical ends, which makes him an experimental philosopher. Perhaps he can be labeled the first experimental philosopher in the 20th century. Interestingly, Benussi has conducted the kind of X-Phi that we have termed intuition-free X-Phi in section 1. We have seen that Benussi's experimental results bring him close to central views of Husserl's phenomenology and that some of his results are of significance for current debates in epistemology.[85]

particularly for Karl Bühler (cf. Münch 1997). For an outstanding recent Husserl-inspired work that demonstrates how philosophical-descriptive and scientific-empirical research can lead to the same results and thereby complement each other, see Madary 2016.

85 At the time this paper was written, the author was a recipient of a DOC Fellowship of the Austrian Academy of Sciences at the Department for Philosophy at the University of Graz. I would like to thank Harald Berger, Johann Marek, and Ulf Höfer for very helpful comments on previous versions of this paper. Also, I am very much indebted to Sonja Rinofner-Kreidl, Harald Wiltsche, and Michael Wallner for many valuable discussions. Special thanks to the editors for helpful remarks and corrections.

References

Albertazzi, Liliana (1995), “Forms of Completion”, in: *Grazer Philosophische Studien* 50, p. 321–340.

Albertazzi, Liliana (2001), “The Legacy of the Graz Psychologists”, in: L. Albertazzi, D. Jacquette, R. Poli (eds.), *The School of Alexius Meinong*, Aldershot: Ashgate, p. 321–345.

Albertazzi, Liliana (2013), “Dissecting Intentionality in the Lab: Meinong’s Theory”, in: *Axiomathes* 23, p. 579–596.

Antonelli, Mauro (1994), *Die experimentelle Analyse des Bewußtseins bei Vittorio Benussi*, Amsterdam: Rodopi.

Antonelli , Mauro / Manotta, Marina (2009), “Meinongs und Benussis Phänomenologie der Wahrnehmung”, in: *Meinong Studies* 3, p. 123–173.

Bealer, George (1992), “The Incoherence of Empiricism”, in: *Proceedings of the Aristotelian Society*, Supplementary Volumes 66, p. 99–143.

Bengson, John (2015), “The Intellectual Given”, in: *Mind* 124, 495, p. 707–760.

Benussi, Vittorio (1913/2002), *Psychologie der Zeitauffassung*, in: *Psychologische Schriften* 2, ed. by Mauro Antonelli, Amsterdam: Rodopi.

Benussi, Vittorio (1906/2002), “Experimentelles über Vorstellungsinadäquatheit”, in: *Psychologische Schriften* 1, edited by Mauro Antonelli, Amsterdam: Rodopi, p. 155–228.

Benussi, Vittorio (1904), “Zur Psychologie des Gestalterfassens (Die Müller-Lyersche Figur)”, in: Alexius Meinong (ed.), *Untersuchungen zur Gegenstandstheorie und Psychologie. Zum zehnjährigen Bestande des Psychologischen Laboratoriums an der Universität Graz*, Leipzig: Barth, p. 303–448.

Berghofer, Philipp (2018), “Why Husserl is a Moderate Foundationalist”, in: *Husserl Studies* 34, 1, p. 1–23.

Boyd, Kenneth / Nagel, Jennifer (2014), “The Reliability of Epistemic Intuitions”, in: E. Machery and E. O’Neill (eds.), *Current Controversies in Experimental Philosophy*, New York: Routledge, p. 109–127.

Boring, Edwin (1929), *A History of Experimental Psychology*, New York: The Century Co.

Cappelen, Herman (2012), *Philosophy Without Intuitions*, Cambridge: Cambridge University Press.

Chudnoff, Elijah (2013), *Intuition*, Oxford: Oxford University Press.

Cokely, Edward / Feltz, Adam (2009), “Individual Differences, Judgment Biases, and Theory-of-Mind: Deconstructing the Intentional Side Effects Asymmetry”, in: *Journal of Research in Personality* 43, p. 18–24.

Cullen, Simon (2010), “Survey-Driven Romanticism”, in: *Review of Philosophy and Psychology* 1, p. 275–296.

Deutsch, Max (2010), “Intuitions, Counter-Examples, and Experimental Philosophy”, in: *Review of Philosophical Psychology* 1, p. 447–460.

Dölling, Evelyn (2001), “Alexius Meinong’s Life and Work”, in: L. Albertazzi, D. Jacquette, R. Poli (eds.), *The School of Alexius Meinong*, Aldershot: Ashgate, p. 49–76.

Erhard, Christopher (2012), “Husserls moderater empirischer Fundamentalismus und das Verhältnis zwischen Phänomenologie, Ontologie und Metaphysik. Kommentar zu Christian Beyer”, in: J. Nida-Rümelin and E. Özmen (eds.), *Welt der Gründe*, Hamburg: Meiner, p. 48–62.

Ghijsen, Harmen (2016), *The Puzzle of Perceptual Justification*, New York: Springer.

Goldman, Alvin (2007), "Philosophical Intuitions: Their Target, Their Source and Their Epistemic Status", in: *Grazer Philosophische Studien* 74, 1, p. 1–26.
Guigon, Ghislain (2005), "Meinong on Magnitudes and Measurement", in: *Meinong Studies* 1, p. 255–296.
Haller, Rudolf (1995/96), "Psychologische Grundlagen der Gegenstandstheorie Meinongs", in: *Brentano Studien* 6, p. 31–41.
Höfer, Ulf (2014), "Über Wahrscheinlichkeit bei Meinong", in: *Meinong Studies* 5, p. 103–159.
Huemer, Michael (2001), *Skepticism and the Veil of Perception*, Lanham MD: Rowman & Littlefield Publishers.
Husserl, Edmund (1973), *Ding und Raum: Vorlesungen 1907*, in: *Husserliana* XVI, ed. by U. Claesges, The Hague: Martinus Nijhoff.
Husserl, Edmund (1984), *Logische Untersuchungen*, Zweiter Band, Erster Teil, in: *Husserliana* XIX/1, ed. by U. Panzer, The Hague: Martinus Nijhoff.
Ichikawa, Jonathan (2014), "Who needs intuitions? Two experimentalist critiques", in: A. Booth and Darrell Rowbottom (eds.), *Intuitions*, Oxford: Oxford University Press, p. 232–256.
Kauppinen, Antti (2007), "The Rise and Fall of Experimental Philosophy", in: *Philosophical Explorations* 10, 2, p. 95–118.
Knobe, Jushua (2016), "Experimental Philosophy Is Cognitive Science", in: J. Sytsma and W. Buckwalter (eds.), *A Companion to Experimental Philosophy*, Malden: Wiley, p. 37–52.
Machery, Edouard et al. (2017), "Gettier Across Cultures", in: *Noûs* 51, 3, p. 645–664.
Madary, Michael (2016), *Visual Phenomenology*, Cambridge: The MIT Press.
Maier, Anneliese (1967), "Das Problem der Evidenz in der Philosophie des 14. Jahrhunderts", in: A. Maier (ed.), *Ausgehendes Mittelalter: Gesammelte Aufsätze zur Geistesgeschichte des 14. Jahrhunderts* II, Roma: Editione di Storia e Letteratura, p. 367–418.
Marek, Johann (2013), "Alexius Meinong", in: Edward N. Zalta (ed.), *The Stanford Encyclopedia of Philosophy* (Fall 2013 Edition), URL = <https://plato.stanford.edu/archives/fall2013/entries/meinong/>.
Meinong, Alexius (1888), *Über den Begriff und Eigenschaften der Empfindungen*, in: *Meinong Gesamtausgabe* I, ed. by R. Haller, R. Kindinger, R. Chisholm, Graz: Akademische Druck- und Verlagsanstalt, p. 109–185.
Meinong, Alexius (1899), *Über Gegenstände höherer Ordnung und deren Verhältnis zur inneren Wahrnehmung*, in: *Meinong Gesamtausgabe* II, ed. by R. Haller, R. Kindinger, R. Chisholm, Graz: Akademische Druck- und Verlagsanstalt, p. 377–480.
Meinong, Alexius (ed.) (1904a), *Untersuchungen zur Gegenstandstheorie und Psychologie*, Leipzig: Barth.
Meinong, Alexius (1904b), "Über Gegenstandstheorie", in: Alexius Meinong (ed.), *Untersuchungen zur Gegenstandstheorie und Psychologie*, Leipzig: Barth, p. 1–50.
Meinong, Alexius (1906), *Über die Erfahrungsgrundlagen unseres Wissens*, in: *Meinong Gesamtausgabe* V, ed. by R. Haller, R. Kindinger, R. Chisholm, Graz: Akademische Druck- und Verlagsanstalt, p. 367–481.
Metzinger, Thomas / Windt, Jennifer (2014), "Die phänomenale Signatur des Wissens: Experimentelle Philosophie des Geistes mit oder ohne Intuitionen?", in: Th. Grundmann, J. Horvath, J. Kipper (eds.), *Die Experimentelle Philosophie in der Diskussion*, Berlin: Suhrkamp, p. 279–321.
Mortensen, Kaija / Nagel, Jennifer (2016), "Armchair-Friendly Experimental Philosophy", in: J. Sytsma and W. Buckwalter (eds.), *A Companion to Experimental Philosophy*, Malden: Wiley, p. 53–70.

Münch, Dieter (1997), "Edmund Husserl und die Würzburger Schule", in: *Brentano Studien* 7, p. 89–122.
Nadelhoffer, Thomas / Nahmias, Eddy (2007), "The Past and Future of Experimental Philosophy", in: *Philosophical Explorations* 10, 2, p. 123–149.
Nado, Jennifer (2014), "The Role of Intuition", in: J. Sytsma (ed.), *Advances in Experimental Philosophy of Mind*, London: Bloomsbury, p. 11–43.
Nichols, Shaun / Knobe, Joshua (2007), "Moral Responsibility and Determinism: The Cognitive Science of Folk Intuitions", in: *Noûs* 41, 4, p. 663–685.
Prinz, Jesse (2008), "Empirical Philosophy and Experimental Philosophy", in: J. Knobe and S. Nichols (eds.), *Experimental Philosophy*, Oxford: Oxford University Press, p. 189–208.
Pinillos, Ángel et al. (2011), "Philosophy's New Challenge: Experiments and Intentional Action", in: *Mind and Language* 26, 1, p. 115–139.
Pust, Joel (2000), *Intuitions as Evidence*, New York: Routledge.
Rinofner-Kreidl, Sonja (2004), "Representationalism and Beyond. A Phenomenological Critique of Thomas Metzinger's Self-Model Theory", in: *Journal of Consciousness Studies* 11, p. 88–108.
Rinofner-Kreidl, Sonja (2000), *Edmund Husserl. Zeitlichkeit und Intentionalität*, Freiburg: Karl Alber.
Russell, Bertrand (1899), "Review of A. Meinong, Über die Bedeutung des Weberschen Gesetzes", in: *Mind* 8, p. 251–256.
Seyedsayamdost, Hamid (2015), "On Normativity and Epistemic Intuitions: Failure of Replication", in: *Episteme* 12, 1, p. 95–116.
Siegel, Susanna (2016), "The Contents of Perception", in: Edward N. Zalta (ed.), The Stanford Encyclopedia of Philosophy (Spring 2016 Edition), URL = <http://plato.stanford.edu/archives/spr2016/entries/perception-contents/>.
Stich, Stephen / Tobia, Kevin (2016), "Experimental Philosophy and the Philosophical Tradition", in: J. Sytsma and W. Buckwalter (eds.), *A Companion to Experimental Philosophy*, Malden: Wiley, p. 5–21.
Tse, Peter et al. (2013), "How Attention Can Alter Appearances", in: L. Albertazzi (ed.), *Handbook of Experimental Phenomenology*, Malden: Wiley, p. 291–315.
Turri, John (2013), "A Conspicuous Art: Putting Gettier to the Test", in: *Philosophers' Imprint* 13, 10, p. 1–16.
Vendrell-Ferran, Íngrid (2009), "Meinongs Philosophie der Gefühle und ihr Einfluss auf die Grazer Schule", in: *Meinong Studies* 3, p. 199–239.
Weinberg, Jonathan (2007), "How to Challenge Intuitions Empirically Without Risking Skepticism", in: *Midwest Studies in Philosophy* 31, p. 318–341.
Weinberg, Jonathan / Nichols, Shaun / Stich, Stephen (2001), "Normativity and Epistemic Intuitions", in: *Philosophical Topics* 29, p. 429–460.
Williamson, Timothy (2007), *The Philosophy of Philosophy*, Oxford: Blackwell.
Williamson, Timothy (2016), "Philosophical Criticism of Experimental Philosophy", in: J. Sytsma and W. Buckwalter (eds.), *A Companion to Experimental Philosophy*, Malden: Wiley, p. 22–36.
Wright, Jennifer (2010), "On Intuitional Stability: The Clear, the Strong, and the Paradigmatic", in: *Cognition* 115, p. 491–503.

Arnaud Dewalque

A Critique of Meinongian Assumptions

Abstract: This article argues that Meinong's analysis of assumption, while exploring the variety of phenomenological primitives in a more promising way than Brentano did, nevertheless fails to adequately account for (i) the noncommittal character of assumptive attitudes and (ii) the difference between assumptive and other neighbouring attitudes. Section 1 outlines an overall framework for the philosophical analysis of assumptions and cognitive attitudes. Section 2 gives an overview of Brentano's analysis of cognitive attitudes and some difficulties thereof. Section 3 offers a critical examination of Meinong's rival analysis. Eventually, Section 4 suggests an alternative route, according to which assumptive attitudes are best analysed in taking entertaining that p as a phenomenological primitive besides affirming that p.

Introduction

It is sometimes held that cognitive states – believing, thinking, understanding, and the like – exhibit a kind of proprietary and distinctive phenomenology, that there is "something it is like" to be in a cognitive state or to undergo a cognitive experience. Over the two last decades or so, this claim has been subject to lively discussions. Typical cases at issue involve what it is like to *understand* the meaning of a spoken or written sentence (as opposed to merely hearing or reading the sentence without understanding it), or what it is like to *know* that a proposition is true. Part of the debate evolves around the alleged cogency of pro and con arguments regarding the *existence* and *reducibility* of cognitive phenomenology. Taking for granted that there is something it is like to understand a sentence, is the understanding experience (say) reducible to an auditory experience plus some kind of mental imagery and/or emotional ingredient?[1] At the same time, the question as to what exactly is the *character* of cognitive experiences and how to give the best description thereof became a hot spot again after a period of neglect.[2]

1 See Bayne / Montague eds. 2014.
2 See Breyer / Gutland eds. 2015

Arnaud Dewalque, University of Liège / University of Luxembourg
a.dewalque@uliege.be

https://doi.org/10.1515/9783110664850-006

Both aspects of the debate have historical antecedents. On the one hand, the question concerning the *existence* and irreducibility of cognitive phenomenology is not unlike the old controversy on "imageless thoughts".[3] On the other, the question concerning the *character* of cognitive experiences is not alien to the phenomenological analyses offered by Brentano and his students. Even though their considerations on judgemental acts are not conducted in terms of phenomenal character, there are reasons to think that classical phenomenologists in the Brentanian tradition simply take it for granted that there is (what is called today) a distinctive phenomenology of judging and neighbouring cognitive attitudes.[4] At stake for Brentano and his followers simply is the *description* of the phenomenological character of subjective experiences, broadly understood.

This paper locates Meinong's theory of assumptions within the theoretical space opened up by the systematic description of cognitive experiences. More pointedly, it focuses on the *character* of assumptive attitudes. The chief issue is how to best describe assumptive attitudes. I argue that Meinong's analysis of assumption, while exploring the variety of phenomenological primitives (*affirming that p* as distinct from *believing that p*) in a more promising way than Brentano did, nevertheless fails to adequately account for (i) the noncommittal character of assumptive attitudes and (ii) the difference between assumptive and other neighbouring attitudes. My plan is as follows. Section 1 outlines an overall framework for the philosophical analysis of cognitive attitudes. Section 2 gives an overview of Brentano's analysis and some difficulties thereof. Section 3 offers a critical examination of Meinong's competing view. And Section 4 suggests an alternative route, according to which assumptive attitudes are best analysed in taking *entertaining that p* as a phenomenological primitive besides *affirming that p*.

1 In Search for Phenomenological Primitives

In this first section I want to suggest that Brentano's and Meinong's analyses of cognitive attitudes are best understood as part and parcel of an overall research programme which aims at discovering phenomenologically primitive notions –

3 Dewalque 2017.

4 See, e.g., Smith 2015, p. 64: "In the phenomenological tradition, since Brentano and Husserl, it seems widely assumed that every conscious experience has a distinctive type of phenomenal or phenomenological character". The reason it is so probably has to do with the fact that, for Brentano and Husserl, intentionality and phenomenality are not disconnected from one another (Kriegel 2013b, p. 2). Yet, I won't say more about that here.

or *phenomenological primitives* for short. This programme rests upon a number of background claims. What follows is an attempt to make those claims explicit and to formulate them in a way that fits the present-day research situation.

First, Brentano's theory of judgement and Meinong's theory of assumption take place within the framework of a "descriptive analysis" of mental phenomena.[5] Their approach to cognitive attitudes is *phenomenological* in the sense that it aims at capturing distinctions that are phenomenally manifest to the subject. It is *descriptive* in the sense that it purports to answer questions of the form *What is P?* And it is *analytic* in the sense that answering such questions requires one to *paraphrase* or recast some notions in terms of others. Historically speaking, this combination of phenomenological description and conceptual analysis is at the heart of Brentano's programme.[6]

Second, central to this programme is the claim that phenomenal distinctions are not exhausted by distinct ways of representing the world, or "contentual distinctions." They first and foremost involve *attitudinal* distinctions. Brentano and Marty emphatically insist that any satisfying description of mental phenomena has to deal with distinct "relational modes of the soul" (*Beziehungsweise der Seele*) or distinct "modes of the intentional relation" (*Modi der intentionalen Beziehung*).[7] The basic assumption is that (i) there are different attitudes toward the same object represented in the same way, and (ii) those attitudes are phenomenally manifest to the subject. For example, believing that today will be a sunny day differs from hoping that today will be a sunny day, and this attitudinal difference is taken to be phenomenally manifest to the subject. If an interpretation of Brentano and Meinong along those lines is correct, then it is probably not unfair to say that their respective analyses of cognitive attitudes aim at capturing what is sometimes called today the *phenomenology of attitude type*.[8]

Third, what are cognitive attitudes? According to a prima facie negative characterisation, cognitive attitudes are mental attitudes which are neither sensory nor algedonic (pain/pleasure) nor emotional.[9] Most of the time, cognitive attitudes may be constructed as attitudes towards a proposition *p*, or *propositional* attitudes. Yet, the question as to whether every cognitive attitude is propositional

5 See., e.g., Brentano 1995a.
6 See, e.g., Dewalque 2014; Röck 2017.
7 See, e.g., Marty 1906.
8 Horgan / Tienson 2002, 522–23.
9 See, e.g., Twardowski 2014, p. 134: "'Thinking' in the broader sense encompasses all mental functions and activities, without exception, which do not belong to the sphere of perception, or to the emotional-volitional sphere, regardless of whether thinking is concrete or abstract."

cannot be settled a priori and should be left unanswered at the outset. For the sake of convenience, however, I'll stick in this article to propositional constructions of the form "φ-ing that *p*." A list of cognitive attitudes would include (but would not be limited to) the following: thinking that *p*, believing that *p*, judging that *p*, entertaining that *p*, supposing that *p*, knowing that *p*, suspecting that *p*, wondering whether *p*, doubting that *p*, etc.

Fourth, I take it that the analytic description of such cognitive attitudes essentially is a matter of how the relevant notions are connected with each other. One basic assumption behind this research programme is that mental verbs like those forming the aforementioned list make up a *web of interconnected notions*. These notions are interconnected in the sense that some of them are likely to be descriptively analysed in terms of some others. This is where the concept of *phenomenological primitive* enters the picture. A *phenomenological primitive* is a notion which is part of the understanding of another, less fundamental notion, and which cannot in turn be understood in terms of other, more fundamental notions. In sum, phenomenological primitives (i) do not admit of any further descriptive analysis – i.e., cannot be paraphrased – and (ii) have a phenomenal import in the sense that they capture a property P whose presence or absence is phenomenally manifest to the subject. At stake is not only the *number* of primitives that are needed in order to account for cognitive attitudes, but also the *nature* thereof and their place in our overall web of phenomenal attitudinal notions. Reduced cognitive attitudes may be said to be *grounded* on more basic notions and, ultimately, on phenomenological primitives.[10] Much of what I have to say in this article will be dedicated to the exploration of such grounding connections.

However sketchy, those indications suffice when it comes to getting a grip on what is involved in the philosophical project of describing cognitive attitudes. At first sight, the agenda for this research programme is to answer these two questions:

Q1 In virtue of what are cognitive attitudes distinguished from one another?

Q2 Which cognitive notions are derived, and which are *conceptually primitive*?

In my view, Meinong's theory of assumptions may be seen as a pioneering contribution to this research programme. Yet, since Meinong's classification of mental phenomena is a critical continuation of Brentano's, I think it advisable to start

---

10 For a similar view, see Kriegel 2015b.

with a brief examination of Brentano's own list of phenomenological primitives. This is the business of the next section.

2 Brentanian Primitives

In the 1874 *Psychology from an Empirical Standpoint* (hereafter PES), Brentano wants to offer a "natural classification" or taxonomy of mental phenomena. This taxonomy "must unite into a single class objects closely related by nature, and it must separate into different classes objects which are relatively distant by nature".[11]

It is common knowledge that, on Brentano's view, the most adequate classification of mental phenomena is into presentations, judgements, and love-and-hate phenomena. This tripartite classification shows a number of interesting characteristics. To begin with, it may be said to be *purely psychological*, in the sense that it is based on (what Brentano takes to be) intrinsic features of mental phenomena rather than extrinsic factors. Now in PES Brentano holds that each type of mental phenomenon has its own, intrinsic way of being intentionally related to something. One way of capturing this idea is to say that, while all mental phenomena are intentional, each fundamental class of mental phenomena is characterised by an "entirely new" mode of intentionality.[12] To give but a trivial example, suppose you think of your favourite ice cream flavour, which happens to be Raspberry Ripple, and then come to *judge that* Raspberry Ripple is your favourite ice cream flavour. Brentano would say that, when moving from *thinking of* (understood as a nominal and noncommittal attitude) to *judging that* (which is committal), you experience an entirely new attitude towards the content of your mental act, to the effect that your favourite ice cream flavour is not only presented but judged about. By contrast with presentations, judgements are committal attitudes. Now an analogous change of intentional attitude occurs, Brentano would go on, when you do not only judge that Raspberry Ripple is your favourite ice cream flavour but also feel an interest for it and desire to eat it. In respect to presentation and judgement, interest is an entirely new way of being intentionally related to something. Feeling an interest for something is committal, too, but not in the same sense as judging: It is a different *attitude* altogether. In sum, Brentano's classification is attitudinal. As George Stout aptly

11 Brentano 1995b, p. 194. I have examined the "natural" character of Brentano's classification elsewhere. See Dewalque 2018 (on which the three following paragraphs are based).
12 See, e.g., Brentano 1995b, 201, etc.

remarks, "differences in the nature of the object are from this point of view irrelevant. Only the attitude or posture of consciousness towards objects is to be taken into account".[13]

A second interesting aspect of Brentano's classification is that his three classes are *ordered* in virtue of one-sided dependence relations obtaining between the related phenomena. Presentations somehow underlie all mental phenomena, while judgements presuppose presentations and interests presuppose presentations and judgements. True, in 1874, Brentano holds the view that emotions or interests are all-pervading in human mental life. Presentations, most of the time if not always, arouse a judgemental and emotional response; they are only a "part" of more complicated mental phenomena. Still, Brentano would argue, it is not impossible for you to have a presentation of a Raspberry Ripple ice cream without judging anything about it, whereas it is impossible for you to judge that Raspberry Ripple is your favourite ice cream flavour without thinking of it or having a presentation of it. By analogy, while it is not impossible to judge that something is the case without experiencing any interest or emotion, it is impossible to experience an interest for something without judging that something is the case.

A third aspect of Brentano's classification as that it is conceived of as being *complete*. However lush one's mental life may be, Brentano argues that, as far as modes of intentionality are concerned, all mental phenomena may be accounted for in terms of presentations, judgements, interests, and combinations thereof. Admittedly, some further distinctions are fairly important, too. Think, for example, of the distinctions between affirmative and negative judgements, self-evident and blind judgements, direct and oblique presentations, love and preference, etc. However, as far as such distinctions correspond to subspecies of the above-mentioned mental species, Brentano takes them to be "non-fundamental".[14] The reason why presentations, judgements, and interests exhaust the number of fundamental classes is that mental phenomena, in Brentano's view, exhibit "*no more and no less* than a threefold fundamental difference in their reference to a content".[15]

Now my present concern is not with Brentano's classification, but with the conceptual relations he advocates when describing judgemental and presenta-

13 Stout 1896, p. 40.
14 See Kriegel 2017, p. 100–101.
15 Brentano 1995b, p. 264; my emphasis.

tional attitudes.[16] Importantly, it is worth emphasising that the notions of judgement and presentation are *not* phenomenologically primitive. Brentano uses the notion of "judgement" whenever something is acknowledged as correct or rejected as incorrect, which suggests that the primitives at play here are *accepting* (or acknowledging) and *rejecting*. And he uses the notion of "presentation" (*Vorstellung*) whenever something "appears" to the subject in the broadest sense of the term or is merely "given in consciousness" without there being any judgemental or emotional attitude involved.[17] In other words, everything one is aware of may be said to be presented in Brentano's sense. This suggests that the primitive notion here is that of *being aware of* something or *being appeared to* by something.

On Brentano's view, judging and having a presentation, albeit non-primitive, nevertheless stand in a *grounding* relation to further cognitive notions. While the notion of judging plays a central role in Brentano's analysis of knowing that *p*, his notion of presentation plays a central role in his analysis of understanding attitudes. As regards the first point, it is plain that Brentano, in line with Aristotle and the Aristotelian tradition, endorses a judgemental account of knowing: "Knowledge is to be found in *judgement* and makes up the perfection of the latter".[18] By contrast, Brentano's account of understanding attitudes is less easy to pin down. Yet, in PES, he clearly argues that understanding some linguistic expression is best analysed in terms of having some presentations: "When I hear and understand a word that names something, I have a presentation of what that word designates; and generally speaking the purpose of such words is to evoke presentations".[19]

Accordingly, Brentano's analysis of cognitive attitudes may be summarised by means of the following paraphrases, which I shall label B0-B3:

B0 S has a presentation of A iff S is appeared to by A.[20]

16 Although presentational attitudes cannot be said to be cognitive in the strict sense of the term, since they are not properly concerned with truth and falsity, I shall include them into the picture, for Brentano maintains that they ground judgemental and understanding attitudes, which clearly have a cognitive import.

17 Brentano 1995b, p. 81, 198, 1956, p. 32.

18 Brentano Ms. EL 80, p. 12.986(4), 1956, p. 23.

19 Brentano 1995b, p. 198.

20 B0 is, so to speak, the "zero level" of cognition, since in Brentano's eyes "being appeared to by A" is noncommittal with respect to A and, therefore, is not truth-assessable. It is, however, prerequisite for any cognitive attitude directed at A.

B1 S judges that A iff S has-a-presentation-of-and-accepts A or S has-a-presentation-of-and-rejects A.

B2 S understands a linguistic expression E iff S has a presentation of what is meant by E.

B3 S knows that *p* iff (i) S judges that *p* and (ii) S's judging that *p* is self-evident to S.

No need to say, further analyses might be added to B0–B3.

Brentano's account provides us with the following answers to our opening questions. With respect to Q2, attitudes which are reductively analysed stand on the left side of the iff-sign, while phenomenal primitives stand on the right side. Therefore, *having a presentation*, *judging* and *knowing* are reducible, non-primitive notions, while *being-appeared-to*, *acknowledging*, *rejecting*, and *self-evidence* are considered phenomenal primitives. With respect to Q1, Brentano holds that those distinctions are *attitudinal only*. Presentation and judgement, for example, clearly are "two entirely different ways of being conscious of an object".[21] Even though it is tempting to introduce some differences at the level of content, such differences are a matter of irrelevance when it comes to analysing the related attitudinal concepts. A presentation is a presentation in exactly the same sense no matter whether it is a presentation of a simple or complex content; and a judgement is a judgement in exactly the same sense no matter whether it has a propositional content (*p*) or a merely nominal content (A). Again, the criterion for such attitudinal distinctions is what Brentano will call, later on, unilateral separability: even if it is not possible for S to judge that *p* without having a presentation of *p*, it is possible for S to have a presentation of something without judging. As shown in B1, having a presentation of something is considered a built-in feature of any judgemental attitude.

Before turning to Meinong's own analyses, I want to briefly review some difficulties with Brentano's analyses. One first difficulty (call it *Difficulty 2.1*) is with phenomenal complexity. On B1, judging is not phenomenally simple, but phenomenally structured. How are we to conceive of such a phenomenal structure? One first option is to say that the new attitude is somehow *added* to the previous one. This interpretation is supported by the claim that judging has presenting as a "psychological prerequisite". Another option is to say that the new attitude is a *modification* of the previous one. This interpretation is supported by the thought

21 Brentano 1995b, p. 201; Stout 1896, p. 40.

that S cannot have a presentation of A and accept or reject A at the same time. This is what Husserl notices. It seems impossible, Husserl argues, to merely have a presentation of something and simultaneously judge about exactly the same thing in exactly the same respect. Presentational attitude, which is noncommittal, and judgemental attitude, which is committal, are incompatible.[22] I think this difficulty is a serious one. For the present purpose, however, there is no need to settle this issue. We can stay neutral on how to best conceive of the phenomenal structure.

A second difficulty (*Difficulty 2.2*) is connected to B1. Interestingly, B1 is disjunctive: Brentano takes it that judging is *either* accepting *or* rejecting. In other words, there are "two attitudinal phenomenal features" of judging rather than one single attitudinal feature of truth-commitment.[23] On the face of it, it is hard to see how this analysis might provide us with a unified understanding of what judging is. Furthermore, it might be objected that acknowledging and rejecting are not phenomenal primitives, indeed that various phenomena are likely to be understood under those notions, which subsequently require some clarification. For example, it may be urged that *judging* is neither *assessing-as-true* nor *assenting to a judgement* – yet, insofar as those three notions (*judging*, *assessing-as-true*, *assenting to a judgement*) are likely to be captured by the notion of *accepting*, the latter turns out to be equivocal (Reinach 1989b; Dewalque 2016). The same holds true, a fortiori, with respect to *affirming* and *believing*, two notions arguably encapsulated within Brentano's notion of acceptance. Therefore, it may be objected that Brentano's analysis is too coarse-grained and that further distinctions are needed.

A third difficulty (*Difficulty 2.3*) with Brentano's account of cognitive attitudes has to do with B2 and, more pointedly, with the understanding of a *sentence*. Let us take for granted that understanding a word (e.g., <tree>, <Albornoz fortress>) amounts to having a presentation of something, as stated in B2. The question arises as to what it is to understand an affirmative or negative sentence (<this tree is an oak>, <Albornoz fortress is the highest point of Urbino>). It cannot be having a presentation corresponding to each word, for there is no presentation for each and every word (e.g., "is", which is a synsemantical term), and it cannot be judging that *p*, otherwise it would be impossible to understand a false proposition without affirming it and thus making an error. Hence the question arises as to what kind of attitude is the understanding of a sentence.

---

22 Husserl 2002, p. 110.
23 Kriegel 2015a, p. 33.

Meinong's theory of assumption may be seen as an attempt of dealing with *Difficulties 2.2* and *2.3*.

3 Meinongian Primitives

It certainly is natural to see Meinong's treatise on assumptions[24] as part of a broader project, namely, that of improving Brentano's classification of mental phenomena. Very roughly, Meinong replaces Brentano's tripartite division by a twofold division into cognitive and emotional phenomena, separates presentations and thoughts as two subclasses of cognitive phenomena, and introduces assumptions as a subclass of thoughts on a par with judgements.[25] Meinong's classification may be illustrated by the following diagram:

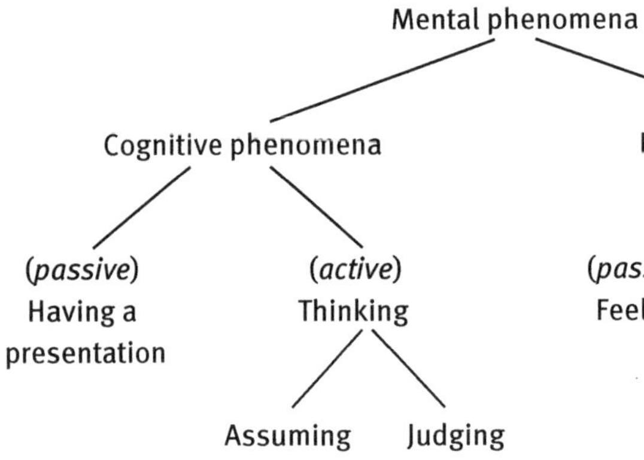

Fig. 1. Meinong's Classification of Mental Phenomena

24 Meinong 1902, 1910, 1977.

25 For a critical discussion, see Marty 1906. Interestingly, Meinong's modifications go hand in hand with the endorsement of a dichotomy method: Mental phenomena are divided into emotional and non-emotional (i.e., cognitive), emotional phenomena are divided into passive (feeling) and active (desires), and so are cognitive phenomena (presentations are passive, thoughts are active, viz. involve an affirmative or negative component). The resulting classification is exposed in Höfler 1894, p. 29 ff., 1930, p. 102 ff. The method of dichotomous division usually is traced back to Aristotle, who explicitly rejected it for biological species, one reason being that dichotomous divisions lead to unnatural groups (Richards 2010, p. 21–22).

In his treatise on assumptions, however, Meinong's goal was not just to demonstrate that assumption is a fundamental attitude between presentation and judgement. It was to show that the notion of assuming, first, is part of the analysans of, and plays the role of a *grounding* notion for, a number of cognitive attitudes, and next, that assuming is a non-primitive attitude which nevertheless cannot be analysed in terms of judgement and presentation. Let's call the first aspect the *Centrality* of assumptive attitudes, and the second one the *Specificity* of assumptive attitudes.

> *Centrality*
> The notion of assuming that p, although it is not itself a phenomenological primitive, is implicit in a number of cognitive attitudes, hence is required for the descriptive analysis of the latters.

> *Specificity*
> Assuming that p, albeit non-primitive, is a sui generis cognitive attitude which is not reducible to having a presentation, judging, or a combination thereof.

Centrality implies that it is not possible to adequately describe or analyse a number of cognitive experiences (analysanda) without including the notion of *assuming that p* in the analysans. One way of reconstructing Meinong's argument for *Centrality* is by means of his notion of "objective" (*Objektiv*) – Meinong's name for a state of affairs. It runs as follows:

P1 If S apprehends an objective, then S judges that *p* or S assumes that *p*.

P2 There are some and, indeed, many states x, y, z, ... such as, when S is in x, y, z, ..., S apprehends an objective and S doesn't judge that *p*.

Therefore,

C there are some and, indeed, many states x, y, z, ... such as, when S is in x, y, z, ..., S assumes that *p*.

Thus reconstructed, the argument is valid. P1 shows a crucial departure from Brentano's analyses. Recall that, for Brentano, the nature of the content (no matter whether nominal or propositional) does not put any constraint whatsoever on the nature of the related attitude: I can have a presentation of <A>, <A which

is B> or <A which is not B>. Similarly, I can accept that <A is B> or I can accept <A> itself, say, in the case of an existential judgement ("A is"). When it comes to determining the attitudinal character of the act, the content of the act is a matter of indifference. P1 breaks with this conception by stating that, whenever what is apprehended is not a single object, but an objective, the apprehending act must be a judgement or an assumption. Meinong's rationale for adopting P1 runs as follows. An objective is a higher-order object ("superiorus") which is founded upon lower-level objects ("inferiora"). Now for founded objects to be apprehended, the corresponding thought has to be actively produced by the mind. Therefore, apprehending an objective requires some judicative or assumptive *activity*. I won't discuss this argument here. Suffice it to note that it presupposes the theory of production which has been elaborated on by Meinong and some of his heirs.[26]

Now Meinong contends that many mental states fall in the scope of P2: meaning that *p*, understanding that *p*, lying, wondering, etc. All those states require one to actively apprehend an objective without judging that *p*.[27] But how are we to account for the so-called "activity" involved here? And in what sense, exactly, does it differ from *judging that p*? Meinong's answer is straightforward: it is a *convictionless affirmation or negation*, and this convictionless affirmation or negation precisely is what is called here an *assumption*. We arrive at this notion of assumption, Meinong remarks, when one notices that the notion of *affirming that p* and the notion of *being convinced that p* do not stand in the determinate-determinable relationship. Affirming that *p* is not a determinate of being convinced that *p*. On the contrary, Meinong argues, *it is quite possible for a subject S to affirm that p without believing or being convinced that p*.[28] This indeed happens very often. For example when S lies to someone, S affirms something she does not believe or denies something she does believe. Similarly, Meinong goes on, when S understands (the thought) that *p*, S does not need to believe that *p*. S merely needs to think that *p*, where *thinking that p* is tantamount to forming the related affirmation or negation "in a neutral way", namely: without forming the related belief or disbelief.

---

26 See, e.g., Höfler 1894; Witasek 1908; etc.

27 Judging is but one kind of "intellectual work" or intellectual "activity" among others. See already Höfler 1894, p. 55 ff.

28 "I would have always thought it obvious that any conviction has to be affirmative or negative, but I would have never expected to find affirmation or negation in any case where conviction was lacking. That this is nevertheless possible and, indeed, anything but rare in its realization – this point, along with the consequences of it, constitutes about the most important thing to be shown by the following discussions" (Meinong 1910, p. 4, 1977, p. 4, 1976, p. 11).

We now are in a position to give a rough picture of Meinong's analysis of cognitive attitudes. On Meinong's view, any nominal content may be apprehended by means of an act of presentation, where an act of presentation of A is a passive apprehension of A. It is passive in the sense that no production of articulated content is involved. By contrast, *thinking that p* is required for a subject S to apprehend the corresponding objective. Now *thinking that p* comes in two forms: either S *judges that p*, to the effect that S *affirms-and-believes that p* (or *denies-and-disbelieves that p*), or S *assumes that p*, to the effect that S *affirms that p without believing that p* (or *denies that p without disbelieving that p*). Accordingly, the 'affirming' or 'denying' component – or, as Meinong says, the yes-no opposition – is common to both judgement and assumption. The difference between the two merely is that, in the case of assumption, the belief component lacks entirely – or, in some cases, is merely reproduced imaginatively.[29] To put it differently, Meinong's analysis rests upon the idea of a *privative contrast* between judgement and assumption: the former does, while the latter does not, exhibit a belief component or an "aspect of conviction" (*Überzeugungsmoment*). As a result, Meinong's analyses suggest that the following equivalences are true:

M0 S has a presentation of A iff S passively apprehends A.

M1 S thinks that *p* (or actively apprehends *p*) iff S judges that *p* or S assumes that *p*.

M2 S judges that *p* iff S affirms-and-believes that *p* or S denies-and-disbelieves that *p*.

M3 S assumes that *p* iff S affirms-without-believing that *p* or S denies-without-disbelieving that *p*.

Moreover, as suggested by *Centrality*, *assuming that p* is a necessary ingredient of further cognitive attitudes. In *Über Annahmen*, Meinong mentions at least the following cases:

---

29 Meinong 1968, p. 333. This does not mean, of course, that in assumption an act of affirmation is combined with an act of disbelief: rather, the affirmation comes alone, without being accompanied by any "conviction" (*Überzeugung*). An assumption, in Meinong's view, is an affirmation (or a negation) which is disconnected from belief (or disbelief).

M4 S means that *p* iff (i) S utters a sentence expressing *p* and (ii) S assumes that *p*.

M5 S understands that *p* iff (i) S has a sensory presentation of a sentence expressing *p* and (ii) S assumes that *p*.[30]

M6 S lies iff (i) S means that *p* and (ii) S disbelieves that *p*.

M7 S wonders whether *p* iff (i) S assumes that *p* and (ii) S desires to know whether *p*.

Let me take stock. Meinong's list of phenomenological non-primitives includes *having a presentation*, *judging*, *assuming*, *meaning*, *understanding*, *lying*, *wondering*, and *thinking*. His list of phenomenological primitives includes *passively apprehending*, *affirming*, *denying*, *believing*, *disbelieving*, and *uttering*. Both lists are open. Moreover, like Brentano, Meinong takes it that judging and assuming presuppose having a presentation. Unlike Brentano, however, he offers a more fine-grained analysis of judging, for he distinguishes between two components that are equally present in judging: the affirming or denying component, and the belief component ("conviction"). This distinction, it may be argued, makes Meinong's analysis more promising than Brentano's, indeed puts him in a better position when it comes to avoiding *Difficulty 2.2*. The belief component and the assertion component (affirmation or negation) are clearly distinguished from one another.

That said, it is a pending question whether M0–M7 adequately account for the phenomena at issue and offer a consistent analysis of cognitive experiences. A full-blown critical discussion should probably consider each equivalence (M0–M7) in turn. However, rather than engaging into detailed case studies, I shall confine myself to mention five general difficulties.

First of all (*Difficulty 3.1*), it may be argued, against M3, that assuming does not show the kind of attitudinal oppositionality judicative attitudes exhibit, namely a *yes-no attitude* which has to be distinguished from the contentual opposition between *p* and *not-p*. If I am not mistaken, there is no sharp distinction between an attitudinal and a contentual opposition in Meinong's analyses. Both sides are thought of as closely connected. This ambivalent status of affirmation

30 See also Russell 1910, 1951, p. 159, 2004, p. 209: "When we can suppose that A loves B, we 'understand the proposition' A loves B. Thus we often understand a proposition in cases where we have not enough knowledge to make a judgement".

and negation becomes still more obvious in Witasek's psychology. When the content of a judgement is "touched" (*berührt*) by the affirming or denying act, Witasek writes, a "quasi-contentual aspect" (*quasi-inhaltliches Moment*) is added to the overall mental phenomenon, to the effect that the content is "complicated with the *yes* or *no*".[31] In other words, with the *attitudinal* oppositionality, some oppositionality is introduced in the *content* of the act of judging. Yet, even if one takes this description for granted, it is questionable that something similar occurs in the case of assumptions. Are there really negative assumptive attitudes? Can you *negatively assume that p*? Does that even make sense? It seems to me that the answer is *no*. Now if we stick to Brentano's (and Meinong's) claim that judging is either accepting or rejecting, the absence of negative assumptions arguably creates a gap between assuming and judging, and speaks for the view that assuming is a non-oppositional attitude like presenting or asserting. Just like a negative assertion is the assertion of something negative, a negative assumption seems to be best construed as the assumption of something negative.[32] This is why Russell, for example, leaned towards an analysis of assumptions in terms of presentations, for presentations, too, are non-assertive attitudes which lack any oppositional character. Be that as it may, it is essential to distinguish between contentual similarity and attitudinal similarity. Russell write: "The greater likeness of assumptions to judgments, we shall conclude, is derived from the identity of their objects and the close similarity of their contents; but *in regard to the act, assumptions are to be classed with presentations*, being merely the presentations of objects of a certain kind".[33] From a merely attitudinal point of view, *assuming that p* is neutral or noncommittal.

A further worry (*Difficulty 3.2*) is with the *incompatibility* between assuming that *p* and judging that *p*. Suppose you are reading and understanding a scientific text but have no idea whether what you are reading is true or not. In that case, Meinong would say, you are merely assuming the content of the text without performing any judgement about it. Then, when you take stance on what you have

31 Witasek 1908, p. 281.

32 After all, this is also what Witasek suggests, willy nilly, when he writes that what assuming that p and judging that p have in common is (i) the content <p> and (ii) the "quasi-contentual aspect" of affirmation and negation (Witasek 1908, p. 311). My proposal is to drop the "quasi-" prefixe and to say that the affirmative or negative aspect present in assumption is contentual through and through, to the effect that a negative assumption actually is the assumption of something negative – full stop.

33 Russell 1904, p. 352, my emphasis. Later on, after Meinong's reply in the second edition of *Über Annahmen*, Russell moved closer to Meinong's position and agreed that assuming or supposing is a sui generis attitude (see Russell 1910, 1951, p. 159, 2004, p. 209).

read, you switch from assumption to judgement. In this scenario, the assumption just is a preliminary attitude – the attitude of *thinking without judging* – you are forced to adopt as long as you are not in the position of taking stance. Now suppose you are asked to consider a counterfactual hypothesis, which you know is false. In that case, the situation is different. It is plain that you can assume something you take to be false. The case of counterfactual hypotheses suggests that S's assuming that *p* is compatible with S's disbelieving that *p*. For instance, you can be asked to assume that <2 + 2 = 5> even if you know (judge with self-evidence) that it is false, hence reject it. Now, on Meinong's analysis, the situation would be this: S assumes, that is *affirms-without-believing that p* and S judges to be false, that is, *denies-and-disbelieves that p*. Thus, S affirms and denies the same propositional content at the same time in the same respect. Now affirming that *p* and denying that *p* are incompatible, indeed contradictory attitudes. Therefore, this way of understanding assumptions fails to account for the compatibility between assuming and judging.[34]

It might be replied that this objection does not take into account the fact that Meinong replaced Brentano's notions of acceptance/rejection by the notions of affirming/denying. The latters presumably refer to a mere attitudinal *component* (not per se a full-blown attitude), which might also be characterised as positing/non-positing something. Now it may be argued that positing/non-positing *p* is quite compatible with the acceptance/rejection of *p*, to the effect that the compatibility problem just vanishes. There is no incompatibility between rejecting-as-incorrect that <2 + 2 = 5> and assuming or positing-without-believing that <2 + 2 = 5>. The case for Meinong's analysis is neatly put by Sébastien Richard: "It is indeed impossible to accept and reject the same matter at the same time and in the same respect, but assumption is not so much about acceptance and rejection than position – the putting of a case".[35]

One way of putting this idea is to conceive of the phenomenal structure of judgements and assumptions in terms of *modification*, rather than in terms of addition: the affirmation (resp., negation) is modified by the presence or absence of belief component, to the effect that the word "affirmation" does not denote the same attitudinal component in *judging that p* and *assuming that p*. However, it is far from being clear whether this move really enables Meinong to resist the objection. The problem comes from the fact that Meinong precisely conceives of positing/non-positing as a *common ingredient* in judgement and assumption. This

34 Marty 1906, p. 21–23.

35 Richard 2017, p. 233. The idea that assuming is best understood as *das Setzen eines Falles* may be traced back to Frege 1891, p. 21. See also Meinong 1977, p. 6, fn. 1.

common affirmative or negative component, which is supposed to be present both in assumption and in judgement, is the source of a serious difficulty, for it blocks any attempt of understanding "affirming" in two different ways.[36] A negative judgement that *p* is supposed to be an act of denying-and-disbelieving that *p*, while an affirmative assumption that *p* is supposed to be an act of affirming-without-believing that *p*, and the two clearly seem mutually exclusive. Again, this suggests that a more neutral attitude than the yes-no attitude is needed if we are to account for such a compatibility.

A similar issue (*Difficulty 3.3.*) is raised as soon as one has to deal with *multiple assumptions*. The situation is described by Brentano in the 1911 appendix to PES: "As Meinong uses [the notion of assumption or supposition], we could often suppose two contradictory things at the same time, as for example when we say, 'Locke says that Descartes is mistaken in teaching that there are innate ideas.' For in this case we would be supposing that someone is mistaken when he teaches that there are innate ideas, and supposing that there are innate ideas, all at the same time".[37] The only way out would be to acknowledge still another sui generis attitude which would be to the assumption what the assumption is to the judgement. I won't say more about that, although this difficulty is maybe less peripheral than it may seem at first sight.

A fourth difficulty (*Difficulty 3.4*) is tied to what I shall call the problem of *converse separability*. Contrary to what Meinong wrotes, it may be objected that affirming-without-believing that *p* is impossible. The idea would be this: Meinong's account takes the phenomenology of affirming wrong, for "every affirmation rests

---

36 Note that this already was the core of Marty's famous objection against *Specificity* – an objection which employs Aristotle's and Brentano's method of *aporia*. Marty's argument runs as follows (Marty 1906, 1908, p. 245): 1) there are only two ways of understanding assumptions as sui generis attitudes: either assumptions and judgements are different genera of mental attitudes or they are species of the same genus; 2) if assumptions and judgements share the same affirmative or negative component, then they cannot possibly be different genera; now 3) they share the same affirmative or negative component (according to Meinong); therefore, 4) they cannot possibly be different genera; 5) if the only difference between them is that assumptions lack of belief component, then they cannot be different species of the same genus, for different species are distinguished from one another by a positive rather than a merely privative contrast; now 6) the only difference between assumptions and judgements (according to Meinong) is that assumptions lack of belief component; therefore, 7) they cannot be different species of the same genus; hence 8) none of the ways of understanding assumptions as sui generis attitudes works. For a more thorough reconstruction of the Marty-Meinong controversy, see Richard 2017.

37 Brentano 1995b, p. 285.

upon a conviction"[38] while "a conviction may exist without any affirmation".[39] For example, on Reinach's account, it seems that a lie doesn't involve an affirmation properly speaking, but only an apparent or *modified* affirmation: at the psychological level, the liar doesn't affirm anything, he merely *simulates* an affirmation. The same might be said of the actor who is playing a role on the theater stage. On Meinong's view, the actor makes affirmations which are assumptions: he *affirms-without-believing*, for example, that he is King Lear. The question at issue is whether this affirmation is an affirmation in the plain sense of the term. My suggestion is that it is best understood as a *simulated* affirmation. But a simulated affirmation is not an affirmation, for "simulated" is here used in a modifying way (in Brentano's sense). If this suggestion is correct, then, pace Meinong, an assumption is not and cannot be an *actual* affirmation. It is only the *appearance* of an affirmation.

Eventually (*Difficulty 3.5*), it may be objected that Meinong's analyses, by relying on *Centrality*, fail to account for a number of descriptive or *phenomenological differences*. For example, it may be argued that there is some phenomenological difference between assuming in the sense of testing a hypothesis and understanding – disregarding the fact that, in the case of understanding, one hears or reads a sentence. For one part, when one performs an assumption in order to reach a certain conclusion, the overall phenomenology arguably involves some interest or conative ingredient, to the effect that one desires to reach a conclusion. Whether a similar conative ingredient is integral to the understanding experience is questionable. Moreover, it is true that, in order to understand a sentence, I don't have to perform a judgement; otherwise, we couldn't understand a sentence that we judge to be false. Understanding that *p* is not judging that *p*. Yet, setting aside Meinong's argument, which conceives of assuming as a way of producing some articulated content, is there any objective reason to analyse understanding that *p* in terms of assuming that *p*? On the face of it, understanding is not the same as assuming. Again, it seems that further distinctions are needed to do justice to the variety of cognitive attitudes.[40]

It seems to me that *Difficulties 2.1-2.5* derive from a common source, namely: the claim (M3) that *assuming that p* is best analysed in terms of *affirming that p* or *denying that p*, as if an assumption would be but an incomplete judgement, a judgement lacking the belief component. We have seen how Meinong arrived at this analysis. The equivalences offered in M0-M3 are not introduced "under

38 Reinach 1989b, p. 125.
39 Reinach 1989a, p. 425.
40 Richard 2017, p. 236–37.

the pressure of the phenomena", so to speak. Rather, they are motivated by Meinong's view that there must be some activity involved in the apprehension of a higher-order object. Even if one agrees with this last claim, however, thinking of this apprehension as constitutively made up of an affirmative or negative *attitudinal component common to judging and assuming* proves highly problematic, for an affirmative or negative attitudinal component necessarily is committal – or it is not attitudinal after all, but merely contentual. Again, it is hard to see how the notions of affirming and denying might be understood in a both attitudinal and noncommittal sense. One main moral to be drawn from all that precisely is that *assuming that p* cannot be construed as a committal attitude like *judging that p*.

To finish, I want to sketch an alternative analysis. Drawing on suggestions made by Broad and – more recently – Kriegel, my proposal is to take *entertaining that p* as a phenomenological primitive besides *affirming that p*.

4 Entertaining as a Phenomenological Primitive

Some authors have suggested that at least some assumptive experiences were to be analysed in terms of *entertaining that p*, to the effect that entertaining should be considered a phenomenological primitive. As far is I know, this suggestion first emerged in a review of Meinong's treatise on assumptions written by Stout's assistant, Charlie D. Broad. Broad writes:

> I cannot help thinking that there are really three different attitudes towards a proposition and that Meinong confounds two of them under the name *Annahme*. These two I would distinguish as supposition and entertainment. It seems to me that entertainment clearly differs from supposition and is presupposed both by it and by judgement. When Meinong insists on the resemblance of *Annahmen* to judgement I think he has suppositions in mind; when he says that every judgement presupposed a corresponding *Annahme* I think he has entertainment in mind. But entertainment as distinct from supposition does not seem to me to differ from having an idea.[41]

Similarly, Twardowski, like Brentano, uses the notion of presentation and insists that what is presented when one understands a sentence is *the judgement expressed by the sentence*:

> When someone maintains that Descartes lived in the 15th century, I perfectly understand this sentence; yet, since I take it that the judgement "Descartes lived in the 15th century" is

41 Broad 1913, p. 92.

> false, it is plain that I cannot perform it. Here, therefore, the understanding of the sentence rests quite obviously on the *presenting* of the judgement denoted by the sentence; performing the judgement thus proves quite irrelevant for the understanding of the sentence.[42]

To this, Meinong would naturally object that an act of judging cannot properly speaking be the object of a *presentation*: "Am I at all able to think of X's judgement that *A* is not *B*, if I do not in the first place somehow think that *A* is not *B*?".[43] Still, even if we grant that some thinking is required here, nothing forces us to endorse (M1) Meinong's analysis of thinking in terms of judging or assuming, let alone (M3) his analysis of assuming in terms of affirming or denying. My own suggestion is that, from a merely attitudinal point of view, a more neutral attitude will do. This is precisely what is offered by Broad's analysis in terms of entertaining.

More recently, Uriah Kriegel proposed a similar analysis. He suggested that entertaining could be conceived of as a phenomenological primitive not unlike Brentano's neutral *Vorstellen*. Importantly, entertaining is an attitude which admits of a number of modifications. I can entertain a proposition in order to decide whether I should believe it or not. But I can also entertain something in a way which is "purely contemplative" and "with no doxastic business in mind".[44] This attitudinal distinction, Kriegel suggests, may be captured by distinguishing between *considering* and *contemplating*. Used in a technical sense, *considering* and *contemplating* are modifications of *entertaining*:

K1 S considers *p* iff (i) S entertains that *p* and (ii) S has the project of figuring out whether *p*.

K2 S contemplates *p* iff (i) S entertains that *p* and (ii) S does not have the project of figuring out whether *p*.

Importantly, those subtle variations do not alter the neutrality which is constitutive of the phenomenology of entertaining. As Kriegel puts it, "entertaining is doxastically neutral on the truth (and goodness) of *p* when performed both in an engaged and in a disengaged mode".[45]

On the face of it, K1 is close to assuming or "supposing". In the case of "supposing", however, there is a crucial difference:

---

42 Twardowski 2017, p. 150.
43 Meinong 1910, p. 118–19, 1977, p. 118–19, 1976, p. 88.
44 Kriegel 2013a, p. 5, 2015b, p. 112.
45 Kriegel 2013a, p. 6.

> Sometimes we entertain that *p* not because we want to know whether *p*, but because we want to know whether *q*, that is, we are interested in the plausibility of some proposition, and our process for figuring out the plausibility of that proposition involves entertaining that *p*. This seems to be what happens when we *suppose* that *p*.[46]

Accordingly, assuming or supposing that *p* would be best analysed as follows:

K3 S supposes that *p* iff (i) S entertains that *p* and (ii) S wants to know what is inferable from *p*.

To be sure, this analysis leaves a number of questions open. It is an open question, for example, whether *entertaining that p* is a common component present in any cognitive attitude (judging that *p*, believing that *p*, and the like). For the present purpose, however, suffice it to say that, on the proposed analysis, *assuming (or supposing) that p* does not involve *affirming that p* or *denying that p* as phenomenal components. I take it that this is a crucial departure from Meinong's analysis. This is not to say, of course, that *affirming that p* and *denying that p* are not phenomenological primitives. Yet, they are not part of the analysans of *assuming that p*.

My own feeling is that such an alternative analysis avoids the above-mentioned difficulties. It avoids *Difficulty 3.1*, since it does not force assumptive attitudes into the yes-no opposition. It also avoids *Difficulty 3.2*, for it makes assumptive attitudes compatible with judgemental attitudes that are directed at the same objective in the same respect. Disregarding the entertaining component they might well be sharing, *assuming that p* and *judging that p* are different attitudes altogether. Similarly, it seems to me that the analysis in terms of entertaining prevents the problem of multiple assumptions (*Difficulty 3.3*) – as well as that of converse separability (*Difficulty 3.4*) – of arising in the first place. Consider, again, the following sentence: "Locke says that Descartes is mistaken in teaching that there are innate ideas." Since, on Meinong's analysis, understanding that *p* involves assuming that *p*, anyone who understands this sentence would have to assume (affirm-without-believing) that there are innate ideas and to assume (affirm-without-believing) that accepting the existence of innate ideas is a mistake. The two affirmations are hardly compatible with one another. If, on the contrary, understanding that *p* is analysed in terms of entertaining that *p*, there is no clash of affirmations involved in the understanding of the above-mentioned sentence. Entertaining that *p* being a noncommittal attitude, there is no contradiction in entertaining (the thought) that there are innate ideas and entertaining (the

---

46 Kriegel 2015b, p. 113.

thought) that the theory of innate ideas is mistaken at the same time. Last but not least, the analysis in terms of entertaining and modifications thereof places us in the position of accounting for fine-grained attitudinal distinctions, including the distinctions between *considering that p*, *contemplating that p* and *assuming that p*.

5 Conclusion

I have argued that Meinong's analysis of assumption fails to adequately account for the noncommittal character of the latter. One main advantage of the alternative approach in terms of *entertaining* is that, with the above-sketched distinctions in hand, we might be able to avoid *Difficulties 3.1-3.4* and to account for some interesting descriptive differences touched upon in *Difficulty 3.5*. Another interesting consequence is that the discrimination of entertaining attitudes implies the reference to some epistemic goals, as obvious from K1-K3. As already suggested by M7, wondering whether *p* involves the desire of knowing whether *p*. This suggests that a full-blown analysis of cognitive attitudes should posit conative attitudes as part of the analysans (wanting, having the project of, etc.). Thus, exploring the connections between cognitive and conative attitudes would be a further route of enquiry.[47]

References

Bayne, Tim / Montague, Michelle eds. (2014), *Cognitive Phenomenology*, Oxford / New York: Oxford University Press.

Brentano, Franz (Ms. EL 80), "Logik. Unpublished Manuscript EL 80. Provisional Online Edition, Franz Brentano Archiv Graz", ed. by Robin D. Rollinger. http://gams.uni-graz.at/archive/objects/context:bag/methods/sdef:Context/get?mode=logik.

Brentano, Franz (1956), *Die Lehre vom richtigen Urteil*, Bern: Francke.

Brentano, Franz (1995a), *Descriptive Psychology*, Engl. transl. by Benito Müller, London: Routledge.

Brentano, Franz (1995b), *Psychology from an Empirical Standpoint*, Engl. transl. by Antos C. Rancurello, D. B. Terrell, and Linda L. McAlister, London: Routledge.

Breyer, Thiemo / Gutland, Christopher eds. (2015), *Phenomenology of Thinking. Philosophical Investigations into the Character of Cognitive Experiences*, New York / London: Routledge.

Broad, Charlie D. (1913), "Critical Notice of A. Meinong, Über Annahmen (Leipzig, 1910)", in: *Mind* 22 (n.s.), p. 90–102.

47 This article is the revised version of a paper presented at the University of Urbino. I'm grateful to the audience there, and especially to Federico Boccaccini, Edoardo Lamedica, Denis Seron and Venanzio Raspa for helpful comments on a previous draft.

Dewalque, Arnaud (2014), "Phénoménologie autrichienne: le programme de Brentano", in: *Austriaca* 78, p. 69–88.
Dewalque, Arnaud (2016), "L'intentionnalité cognitive et ses modes : Reinach Critique de Brentano", in: *Philosophie* 128, 1, p. 20–33.
Dewalque, Arnaud (2017), "Külpe on Cognitive Attitudes", in: *Discipline Filosofiche* 27, 2, p. 157-176.
Dewalque, Arnaud (2018), "Natural Classes in Brentano's Psychology", in: *Brentano Studien* 16, p. 107–138.
Frege, Gottlob (1891), *Funktion und Begriff*, Jena: Pohle.
Höfler, Alois (1894), *Psychische Arbeit*, Hamburg / Leipzig: Voss.
Höfler, Alois (1930), *Psychologie*, 2nd ed., vol. 1., Wien / Leipzig: Hölder – Pichler – Tempsky.
Horgan, Terence E. / Tienson, John (2002), "The Intentionality of Phenomenology and the Phenomenology of Intentionality", in: *Philosophy of Mind. Classical and Contemporary Readings*, ed. by David J. Chalmers, Oxford / New York: Oxford University Press, p. 520–33.
Husserl, Edmund (2002), *Urteilstheorie. Vorlesung 1905* (= Husserliana Materialien, Bd. 5), ed. by Elisabeth Schuhmann, Dordrecht: Kluwer.
Kriegel, Uriah (2013a), "Entertaining as a Propositional Attitude: A Non-Reductive Characterization", in: *American Philosophical Quarterly* 50, 1, p. 1–22.
Kriegel, Uriah (2013b), "The Phenomenal Intentionality Research Program", in: *Phenomenal Intentionality*, ed. by U. Kriegel, Oxford: Oxford University Press, p. 1–26.
Kriegel, Uriah (2015a), "The Character of Cognitive Phenomenology", in: *Phenomenology of Thinking. Philosophical Investigations Into the Character of Cognitive Experiences*, ed. by Thiemo Breyer and Christopher Gutland, New York / London: Routledge, p. 25–43.
Kriegel, Uriah (2015b), *The Varieties of Consciousness*, Oxford: Oxford University Press.
Marty, Anton (1906), "Über Annahmen (Ein Kritischer Beitrag Zur Psychologie, Namentlich Der Deskriptiven)", in: *Zeitschrift für Psychologie und Physiologie der Sinnesorgane* 40, p. 1–54.
Marty, Anton (1908), *Untersuchungen zur Grundlegung der allgemeinen Grammatik und Sprachphilosophie*, Halle: Niemeyer.
Meinong, Alexius (1902), *Über Annahmen*, 1st ed., Leipzig: Barth.
Meinong, Alexius (1910), *Über Annahmen*, 2nd ed., Leipzig: Barth.
Meinong, Alexius (1968), "Über emotionale Präsentation", in: *Gesamtausgabe*, 3: *Abhandlungen zur Werttheorie*, Graz: Akademische Druck- u. Verlagsanstalt, p. 283-465.
Meinong, Alexius (1976), *On Assumptions*, Engl. transl. by J. Heanue, Berkeley: University of California Press.
Meinong, Alexius (1977), *Über Annahmen*, in: *Gesamtausgabe*, 4, Graz: Akademische Druck- u. Verlagsanstalt, p. iii–xxv, 1–384.
Reinach, Adolf (1989a), "Einleitung in die Philosophie (1913)", in: *Sämtliche Werke*, 1, München: Philosophia Verlag, p. 369–514.
Reinach, Adolf (1989b), "Zur Theorie des negativen Urteils (1911)", in: *Sämtliche Werke*, 1, München: Philosophia Verlag, p. 95–140.
Richard, Sébastien (2017), "Marty against Meinong on Assumptions", in: *Mind and Language – On the Philosophy of Anton Marty*, ed. by Guillaume Fréchette and Hamid Taieb, Berlin: De Gruyter, p. 219–240.
Röck, Tina (2017), "Brentano's Methodology as a Path through the Divide: On Combining Phenomenological Descriptions and Logical Analysis", *Axiomathes* 27, 5, p. 475–489.

Russell, Bertrand (1904), "Meinong's Theory of Complexes and Assumptions (II)", *Mind* 13, 51 (n.s.), p. 336–354.
Russell, Bertrand (1910), "Knowledge by Acquaintance and Knowledge by Description", *Proceedings of the Aristotelian Society* 11, p. 108–128.
Russell, Bertrand (1951), "Knowledge by Acquaintance and Knowledge by Description", in: *Mysticism and Logic*, 2nd ed., Totowa: Barnes & Noble, p. 152–167.
Russell, Bertrand (2004), *Mysticism and Logic*, 3rd ed., London: Routledge.
Smith, David Woodruff (2015), "Phenomenally Thinking about this Individual", in: *Phenomenology of Thinking. Philosophical Investigations into the Character of Cognitive Experiences*, ed. by Thiemo Breyer and Christopher Gutland, New York / London: Routledge, p. 62–76.
Stout, George (1896), *Analytic Psychology*, vol. 1, London: Swan Sonnenschein & Co.
Twardowski, Kasimierz (2014), *On Prejudices, Judgements and Other Topics in Philosophy*, ed. by Anna Brożek and Jacek Jadacki, Amsterdam: Rodopi.
Twardowski, Kasimierz (2017), "Über begriffliche Vorstellungen", in: *Gesammelte Deutsche Werke*, ed. by Anna Brożek, Jacek Jadacki, and Friedrich Stadler, Cham: Springer, p. 145–164.
Witasek, Stefan (1908), *Grundlinien der Psychologie*, Leipzig: Meiner.

Cristina Travanini

Translating *Über Annahmen*: Questions and Perspectives

"Saying *almost* the same thing."
U. Eco

Abstract: In this article, I intend to address some questions that have arisen while translating Alexius Meinong's second edition of *Über Annahmen* from German into Italian, and some research perspectives that might be significant for current philosophical debates. I start with some remarks on the practice of translating a philosophical work before outlining the editorial history of *Über Annahmen* and offering an overview of Meinong's account of assumptions. Finally, I consider some research perspectives that might benefit from a direct confrontation with Meinong's work. In the background, I try to answer the question: why still engage with this book, and why invest time and energy into working on it?

1 Translation as "Negotiation"

"Negotiation" is the concept that allows us to best understand the translation practice. With that, I mean a negotiation between the *literary sense* and the *general sense* of a sentence, a negotiation that often removes a certain amount of effects implied by an original term.

In this sense, when we translate, we never say exactly the same thing, but *almost* the same thing, in an approximation to the original meaning, which can rarely be perfectly achieved.[1] Given a nuclear content that the translation of a word needs to express, the translator chooses the term that best conveys the same content in the corresponding language. In case of partial loss of the original meaning, the translator always tries to compensate, for example by using periphrasis.

While translating *Über Annahmen*, I kept a few remarks on the translation practice in mind. When the text was particularly obscure, I respected ambiguity

1 Eco 2003.

Cristina Travanini, University of Parma
cristina.travanini@libero.it

https://doi.org/10.1515/9783110664850-007

and did not force it into clarity. I also respected the concept of "reversibility", according to which it is always possible to return from the translated text to the original one. I was, however, well aware of the fact that reversibility must also be negotiated. Since a translation might be target-oriented or source-oriented, I negotiated the orientation needed sentence by sentence, following the belief that a faithful translation is not an exact one, but the most honest in terms of negotiations.[2]

As a few scholars have noted, Meinong can be considered the philosopher of continuity and graduality.[3] In his theory of knowledge, conceived as a constant approximation of complete objects through intending incomplete ones, we see a challenge that is similar to that of the translator: the attempt of reinstituting a sense of unity and continuity.

In Table 1, I list some of the words that required a particularly careful examination to illustrate some of the decisions I made. The last words (*werten-werthalten-bewerten*) perfectly exemplify the concept of translation as negotiation: as you can see, in Italian, it was impossible to find an immediate translation of the German verbs, and I had to rely on a more complicated, and less elegant, periphrasis.

Tab. 1

DEUTSCH	ENGLISH [1983]	ITALIAN
Objektiv	objective	oggettivo
Objekt	objectum	obiectum
Außersein	absistence	extra-essere
Vorstellung	representation	rappresentazione
Präsentation	presentation	presentazione
Wirklichkeit	reality	attualità
werten	valuation	valutare sulla base di assunzioni
werthalten	valuing	valutare sulla base di giudizi
bewerten	evaluation	valutare intellettuale

2 Eco 1995.
3 Raspa 2005.

2 Main Changes Between the Two Editions

The differences between the first edition (1902) of *Über Annahmen* and the second (1910) reflect the theoretical development of Meinong's theory of object. In 1904, Meinong published his programmatic text, *Theory of Object*, whose influence can be found in this second edition. Even though *Über Annahmen* was originally conceived of as a work on psychology, concerned with mental facts, the second edition presents some changes that come directly from Meinong's research on object theory.

Tab. 2

Chapter	I Edition (1902)	II Edition (1910)
1	Erste Aufstellungen	Erste Aufstellungen
2	Zur Frage nach den charakteristischen Leistungen des Satzes	Zur Frage nach den charakteristischen Leistungen des Satzes
3	Die nächstliegenden Annahmefälle	Das Objektiv (1902, ch. 7)
4	Die Annahmeschlüsse	Die nächstliegenden Annahmefälle (1902, ch. 3)
5	Zur Gegenständlichkeit des Psychischen	Das Ojektiv und die Annahmen (1902, ch. 7)
6	Erfassen von Gegenständen höherer Ordnung	Annahmen bei Operationen an Objektiven (1902, ch. 5)
7	Das Objektiv	Annahmen beim Erfassen des Präsentierten. Das Meinen (1902, ch. 5)
8	Begehrungs- und Wertpsychologie	Annahmen bei Complexen. Weiteres über das Meinen (1902, ch. 6)
9	Ergebnisse. Bausteine zu einer Psychologie der Annahmen	Begehrungs- und Wertpsychologie
10		Ergebnisse. Bausteine zu einer Psychologie der Annahmen

The most apparent difference between the two editions involves the notion of objective, whose investigation in the first edition was limited to Chapter 7, and has been extended in 1910 to Chapters 3 and 5. Next, in the second edition, Meinong rephrases the question of the objectivity of representations: he rejects the "prejudice in favor of the actual", which was still present in the first edition, and replaces the "being-view" with the "*Aussersein*-view" (Chapter 7 of the second edition). The three introductory chapters of the first edition remain relatively untouched, as well as Chapter 8 – on the psychology of emotions and desires – of the first edition, which becomes Chapter 9 in the second. Additionally, Meinong added a new paragraph (§ 64), "On the possibility of assumption", to Chapter 10 – otherwise largely unchanged – which makes explicit his awareness of the innovative approach of his theory, and consequently of the potential resistance against it within the contemporary philosophical and psychological scene.

In other words, the second edition shows that the analysis of the objective gains much more importance, that the prejudice in favor of the actual is explicitly rejected, and that a new dimension of extra-being (*Außersein*) comes into play.

In *Über Annahmen*, Meinong has an explicit, twofold aim: first, he wants to characterize a new class of experiences, namely assumptions, clarifying their *position* and *function* in relation to their psychological setting; and second, he outlines his theory of intentionality.

3 Mental Setting and Function of Assumptions

As is well known, Meinong studied in Vienna with Franz Brentano from 1875 to 1878. In his *Psychology from an Empirical Standpoint*, Brentano recognized three kinds of mental acts: representations,[4] judgments, and emotions [*Gemütsbewegungen*]. From his mentor's theory of intentionality, Meinong took some important theses that he would never reject: that every mental act is defined by a particular direction toward an object; that representations (*Vorstellungen*) are the most basic psychological act, since they are necessarily presupposed by thoughts, assumptions, and emotions; and finally, that emotions constitute a specific form of intentional act, which is responsible for the apprehension of values.

Meinong identifies four experiences through which an object can be apprehended: representations [*Vorstellungen*], thoughts [*Gedanken*], feelings [*Gefühle*], and desires [*Begehrungen*]. Thoughts include not only judgments [*Urteile*], but also assumptions [*Annahmen*].[5]

4 Throughout this contribution, Vorstellung will be translated by means of the word "representation", in order to avoid confusion with the term Präsentation.
5 Meinong [1983], p. 11.

Meinong uses the concept of representation more narrowly than Brentano. For Brentano, to be represented means to appear in consciousness [*Bewußtsein*] and is closer to what Meinong calls *Präsentation* (translated as "*Presentation*" / "*presentazione*").[6]

Judgments are radically different from representations: since they always entail a certain element of conviction, they are necessarily expressed by sentences that are either affirmative or negative. Furthermore, they are *active* experiences: indeed, to judge is always *to do* something.

Between representations and judgments, Meinong identifies an "intermediate" experience, which corresponds to the domain of *assumptions*, defined as "convictionless" affirmations and negations,[7] and "judgments without conviction".[8] In this sense, an assumption is "more than a mere representation and less than a judgment";[9] it is similar to a judgment in that both have propositional content and apprehend objectives. Like judgments, assumptions are active, dependent experience – necessarily based on preliminary representations – while representations and feelings are passive.[10] But an assumption differs from a judgment in that it does not entail any belief in the being of the apprehended object.

Over the course of the book, this account of assumptions undergoes several changes: in particular, in the last chapter, the assumption becomes a sort of "surrogate" of the judgment that is involved whenever it is a matter of indifference whether an object exists or not, and whether an objective is factual or not.[11] Here, the assumption is also defined as "a sort of limit-case of the judgment, characterized by the zero-value of the strength of conviction".[12] Meinong now proposes considering assumption and judgment together under the heading of "thoughts", a new class of experiences to which both assumptions and judgments belong. In this last chapter, the assumption seems to be a stage preliminary to the judgment rather than something different.

On Meinong's view, there are three main cases in which assumptions appear in the processes of the mind:

A. Assumptions are necessarily involved in arts, play, and lies (as in Chapter 4 – on *explicit* assumptions), in which again the apprehended objectives are

---

6 On the concept of Präsentation, see Meinong 1910, § 4; Meinong 1917.
7 Meinong [1983], p. 11.
8 Meinong [1983], p. 263.
9 Meinong [1983], p. 262.
10 Meinong [1983], p. 243.
11 Meinong [1983], p. 256.
12 Meinong [1983], p. 245.

indifferent to any consideration of their being. Assumptions become necessary components of our attitude towards fiction not only from an intellectual point of view but also from an emotional one.

B. Assumptions are fundamental components of basic intellectual operations. They are concerned with many *operations on objectives*, in hypothetical or disjunctive judgment, as well as in inferences "with suspended premises", as described in Chapter 6. As a rule, Meinong states that, whenever an objective is apprehended without claiming the existence (or non-existence) of the object at issue, an assumption is involved.

C. Assumptions are also psychological prerequisites for many extra-intellectual activities. In particular, all *desires* require the intervention of assumptions, since their very nature precludes the factuality of the objectives involved; we may only desire what does not exist *yet*. This means that judgments are excluded.

In what follows, I identify some current research perspectives that might take advantage of a direct confrontation with Meinong's theses in *Über Annahmen*. I focus here on the current debate on the rationality of emotions and on imagination.

4 Research Perspectives

Analytic philosophers have rediscovered Meinong's thought in recent decades; in particular, his account of non-existent objects has been exercising significant influence over a wide range of fields, from logic to ontology. Instead, Meinong's account of intentionality has received less attention, especially in its emotional declination.13 In what follows, I argue that Meinong's account of emotions – outlined in *Über Annahmen* – anticipates some issues of the contemporary philosophical debate, namely the "rationality" of emotions and the concept of "quasi-emotion".

As a necessary premise, I must make explicit that Meinong does not distinguish between emotion and feeling in the same way that contemporary philosophy of emotions does.[14] "Emotion" is for Meinong a wider category that includes

---

13 Christine Tappolet has done pioneering work on Meinong's account of emotions. More recently Federico Lauria and Olivier Massin have traced their views on desires back to Meinong's emotional intentionality (see Tappolet 2000; Lauria 2017; Mossin 2017).

14 In the contemporary debate, feelings are defined as bodily sensations, disturbances and agitations, which determine the phenomenology of emotions. Anger, for example, "may lend itself to a description in terms of a configuration of sensations caused inter alia by the following elements: an accelerating heart rate, quickened breathing, an increased blood pressure, a rush of adrenaline" (Deonna / Teroni 2012, p. 2).

feelings, which are directed toward what we generally define as values – the good, the beautiful, the true, the pleasant – and *desires*, which are directed toward purposes and duties.

Here, I take into consideration two arguments discussed in *Über Annahmen*:

I. The "quasi-intellectual", "rational" functioning of emotions, which would be explicitly outlined by Meinong in 1917, is already apparent in *Über Annahmen* (in Chapter 9) and is well exemplified by the controversy with Ehrenfels on the nature of values.
II. The notion of "imaginative feeling", which is close to the notion of "quasi-emotions" currently in use in philosophical debates about fiction.

5 The Debate on Emotions: Rationality and Objectivity

Meinong's work presents an analytic examination of what an emotion is. Emotions are peculiar experiences, characterized by a specific, bodily-felt *Erlebnis* and a certain direction toward an object. Nonetheless, *emotional* intentionality is, according to Meinong, weaker than *intellectual* intentionality. Feelings and desires cannot apprehend the object "in a complete way [*fertig*]";[15] they require a preliminary representation of the object in question and an underlying assumption or judgment about it – hence the intervention of thoughts (judgments or assumptions).

> As far as is presently known, feelings and desires lack the capacity to approach objects immediately, no matter whether these objects are objecta or objectives. Hence, whenever an objective is accosted by an emotional experience – or more briefly, whenever an objective is emotionally approached – this at the same time signalizes an intellectual approach.[16]

To apprehend – to know – an object always requires an intellectual operation; though emotional experiences have a "presentational" ability, they cannot achieve knowledge of the object. Meinong distinguishes between a *potential* apprehension [*erfassen*] of the object, easily accomplished by simple representations, and an *actual* intending [*meinen*] the object, which only thoughts are able to achieve.[17] The emotional presentation requires, therefore, a double-layered preliminary intellectual intervention: a representation of the object *and* a presuppositional thought that actually *intends* the object in question.

---

15 Meinong 1917, p. 403.
16 Meinong [1983], p. 122.
17 Meinong [1983], p. 164.

Given this connection between emotions and preliminary thoughts, Meinong emphasizes that emotions also must respect epistemic standards of correctness; only emotions grounded on *justified* judgments might be defined in turn as "*justified*".

> Emotions can be defined as justified if the judgments which attribute their object to their presuppositional objects are justified.[18]

To define an emotion as "correct" or "incorrect", "appropriate" or "inappropriate", means that the emotional and conative domain is no longer necessarily subject to the irrationality of individual inclinations. Instead, it is also possible to identify logical dynamics in the realm of emotions, and to define an emotion as "correct" or "appropriate" in connection with a certain object.

In *On Emotional Presentation*, Meinong wonders whether it is possible to find anything analogous to intellectual evidence in the field of emotions. While denying the possibility of immediate evidence, Meinong defends a form of *mediate* evidence for the emotional sphere, since there are rules of inference [*Schlußgesetze*] that emotions must respect. For instance, if I assign a value to a given object *x*, it is *reasonable* to assign a value also to the *cause* of *x*.[19]

Moreover, we can also consider the case of opposite feelings [*Gegengefühle*], where, for example, one cannot rejoice at the same time for the existence *and* for the non-existence of something.[20] This means that the emotional domain also reflects *a priori* principles of rational consistency:[21] indeed, we might recognize the *a priori validity* of certain connections among emotions, and the presence of logical dynamics within the emotional sphere – something analogous to the consequentiality between premises and conclusions that we find in the intellectual domain.

The controversy with Ehrenfels, examined in Chapter 9 of *Über Annahmen*, also constitutes a way to demonstrate the *rational consistency* of emotions, showing that feelings are "logical antecedents" to desires. In section 53, Meinong challenges Christian von Ehrenfels' theory of value, holding that it is against the *rules of intentionality*, and hence "irrational" to define values in terms of "being desirable".

Beginning in his first work on ethics, *Psychologisch-ethische Untersuchungen zur Wert-Theorie* (1894), Meinong upheld the evidence of the role played by feelings in value experience. "An object has value", he claims, "as far as it has the

18 Meinong [1972], p. 123.
19 Meinong 1917, p. 412.
20 Meinong 1917, p. 410.
21 Meinong 1917, p. 412.

ability to provide an effective basis for a value-feeling [*Wertgefühl*]".[22] In *Über Annahmen* (1902), Meinong develops and modifies his value theory while refuting Christian von Ehrenfels' rival position, whose definition of value rests on desires. For Ehrenfels, an object presents a certain value as far as it is desired, or at least as far as it *could* be desired. In this sense, according to Ehrenfels, desire is preliminary to value.[23]

Meinong rejects this definition of value as "desirability" [*Begehrbarkeit*] in a twofold way. First, from an empirical perspective, he observes that what we feel as value does not always become a motivation to act, and therefore it does not always motivate a desire. Depressed people, for instance, sink into sadness and dissatisfaction although they, too, value happiness. This scenario proves that our definition of value is primarily based on feelings, and that desires come after, once we have already had a value-experience. Vice versa, we might also attach value to something that we *cannot* desire, for instance, to something in the past, which is not "desirable" anymore, since it already exists.[24]

Second, it may be possible to refute Ehrenfels' position *a priori* as well, while analyzing the dynamics of the emotional acts involved. For Ehrenfels, desires are preliminary to feelings since we desire something in order to feel a certain pleasure. Ultimately, pleasure becomes a value in itself. – In other words, he defines a value as the ability of an object to produce pleasure, becoming therefore *desirable*.

Given that only actual things may produce pleasure, Meinong objects that desiring something does not produce pleasure by itself, since (*ex definitione*) what I desire does not exist yet. According to Meinong, value is rather the "logical antecedent" [*das logisch Frühere*][25] of any desire.

> When someone says that he desires *A* because it is enjoyable to him, by this "it" he does not mean the desiring (the *act*). What he means is most certainly in some manner *that* which is desired, or *that* which can be desired. [...] But right there lies the old difficulty: what is desired cannot yet give me pleasure, since as that which can be desired it is not yet realized.[26]

Emotions are therefore examined within the logical framework of an intentional structure: they are acts with a certain content, directed toward a specific object.

---

22 Meinong 1894, p. 37.

23 Meinong takes here into consideration Ehrenfels' *System der Werttheorie* (1897). For a thorough analysis of Ehrenfels' position on values, see Reicher 2009.

24 Meinong [1983], p. 233.

25 Meinong 1923: 521.

26 Meinong [1983], p. 219.

In this sense, Meinong suggests that not only representations and thoughts, but also feelings and desires, are "objective", given that they necessarily have a certain direction toward an object. Admittedly, an "objective" act is always bound to a certain *subject* who feels it; at the same time, this concept of "subjective" does not imply in itself any sort of relativistic consequence.

The current debate on emotions in the philosophy of mind broadly agrees on three main theses, which confirm the results of Meinong's investigations:[27]

I. *Phenomenology*. Any emotion has a certain phenomenological quality because it is *felt* in a specific way. Any emotion is related to a feeling of pleasure or displeasure, an emotional opposition that corresponds to the intellectual one between affirmation and negation: joy and happiness are positive states, while pain and fear are negative ones.
II. *Intentionality*. Emotions are *intentional* states since they are directed toward peculiar objects. A specific, formal object characterizes each kind of emotion. Depending on the intentional object, emotions might be distinguished from other affective phenomena, such as desires and moods.
III. *Epistemology*. Emotions also must respect epistemic standards of correctness; only emotions grounded on *justified* judgments might be defined in turn as "*justified*". Indeed, emotions are always states for which we seek reasons: Why are you angry? Why exactly are you scared? Emotions raise normative questions about the extent to which they can be said to be rational or can contribute to rationality.

6 Towards a Phenomenology of Imagination

The second point I would like to focus on is the emotional attitude toward fiction. Since an assumption is a judgment without belief, the assumption is thus linked to the human ability to *imagine* something that does not obtain – that is, something fictional or non-existent. The concept of assumption explains our apprehension of fictional objects, insofar as we make assertions about them without affirming their existence.

Now, assumptions play a fundamental role not only in the intellectual attitude towards fiction but also in our emotional appreciation of it, since they are the necessary prerequisite for the emergence of feelings and desires about fic-

27 Deonna / Teroni 2012, p. 1-13.

tional discourse. In Chapter 9 of *Über Annahmen,* Meinong identifies peculiar imaginative feelings and desires that are responsible for the emotional reaction to fiction.

The problem of feelings toward the fictional discourse can be summarized like this:

> What are they, really – the "fear" and the "compassion" or whatever it may be that tragedy has the task of "evoking"? A fear in which one is basically not afraid at all, a compassion which, more closely regarded, is seen not to involve any sense of pain whatsoever – are they yet "feelings" as the psychologist is in the first instance accustomed to regard them?[28]

Meinong calls these feeling-like experiences "quasi-feelings" [*Quasi-Gefühle*],[29] which are based on imaginative representations and assumptions. Like serious feelings, quasi-feelings partake of the opposition of pleasure and displeasure, but they are grounded neither on perceptive representations nor on judgments. The spectator in a theater actually experiences something that is *not* literally joy or sorrow, fear or hope, but *something similar.* These "quasi-feelings" are rather "shadowy"[30] in comparison with serious ones: they are characterized by "fleetingness" [*Flüchtigkeit*], a certain absence of "color" [*Bläße*], and they have no connection to motivation.31

Obviously, imaginative feelings are always connected to imagination, which can be conceived not only as fiction, but also as a recollection of something from the past. The quasi-feelings, i.e. the imaginative feelings based on assumptions, constitute a *subclass* of imaginative feelings. As Table 3 shows, not all imaginative feelings have assumptions as their psychological prerequisites. For example, when I recollect a past sensory feeling, such as a toothache, I only need an imaginative representation, not an assumption. Or, if I remember someone's death, I need an imaginative representation and a judgment, not an assumption. Conversely, not all assumptions bring imaginative feelings along with

---

28 Meinong [1983], p. 221.

29 Meinong [1983], § 58.

30 Meinong 1923, p. 128-129.

31 Meinong [1983], p. 269. This idea of quasi-emotions was widely investigated by the Graz School: while Stephan Witasek rejected it, Carl Lange, Moritz Geiger and Oswald Külpe defended it. In the analytic debate on aesthetics, Gilbert Ryle, Anthony Kenny and Kendall Walton supported this idea independently of the Graz School; only Kevin Mulligan makes an explicit reference to the Austrian philosophy. In this regard, I agree with Ingrid Vendrell Ferran, when she advocates a closer connection between the Graz School and the current debate on quasi-emotion. See Mulligan 1998; Vendrell Ferran 2010.

them. For instance, aesthetic feelings are based on assumptions but are serious (not imaginative), as in the case of artistic appreciation or delight for a work of art.

Tab. 3

	Experiences ***Erlebnisse***	**Psychological Prerequisites**	**Example**
Serious Feelings ***Ernstgefühle***	Sensory Feelings *Sinnliche Gefühle*	Representation	Toothache
	Aesthetic Feelings *Ästhetische Gefühle*	Representation Assumption	Delight in a work of art
	Value-Feelings *Wertgefühle*	Representation Judgement	Joy for a friend's achievement
	Knowledge-Feelings *Wissensgefühle*	Representation Judgement	Joy for knowing something
Shadowy Feelings ***Schattenhafte Gefühle***	Imaginative Feelings *Phantasiegefühle*	Imaginative Representation *Phantasievorstellung*: (Imaginative Perception *Phantasiewahrnehmung)*	Fear of a ghost
		Imaginative Representation: (Memory *Erinnerung)*	Fear of a past storm
		Imaginative Representation (Memory) Judgement	Sadness for a friend's past grief
	Quasi-Feeling *Quasi–Gefühl*	Imaginative Representation Assumption	Sadness for Anna Karenina's death

There are three main cases in which mental processes involve imagination and assumptions:

I. Imagination is involved in basic intellectual operations, such as in hypothetical or disjunctive reasoning, as well as in inferences with suspended premises. As a rule, Meinong states that whenever a state of affairs is apprehended without claiming the existence or the non-existence of the object involved, an assumption intervenes and an imagining process takes place.

II. Imagination is also necessarily involved in human *desires*; since their very nature precludes the factuality of the objects involved, we have to imagine how it would be *if* the object we desire existed. Indeed, we may only desire what does not exist *yet*.

III. Imagination is connected to arts, play, and lies, or whatever situation in which the apprehended state of affairs is indifferent to any consideration of its being or nonbeing. In fictional discourse, the imagination process also affects our emotional response toward fictional entities, such as the feelings experienced by the spectators in a theatre or by *Anna Karenina*'s readers.

In the end, what we find in Meinong's views is a sophisticated account of imagination, connected not only to intellectual life but also to our emotional attitude towards non-existent entities and states of affairs. Meinong suggests that imagination deals with a manifold phenomenon that plays out on the center stage of consciousness, and this accounts for the complexity of human mental life: when we make thought experiments and hypotheses in both ordinary and scientific reasoning, when we remember or anticipate something, as well as in desires and feelings, imagination plays a fundamental role. Anytime we are interested in what is *possible* over what is real, imagination comes into play.

We can observe a two-stage character in each class of mental facts. Imaginative representations, imaginative judgments (assumptions), imaginative feelings, and imaginative desires, which constitute the lowest stage of all four classes of mental experiences, appear as a "homogenous whole" (p. 271) because of the connection with the concept of imagination. Imagination then becomes a transversal phenomenon that cuts across the intellectual *and* the emotional dimension.

By acknowledging the significance and value of the imagination process, Meinong overcomes the dichotomy between imagination and perception, emotion and thought, which characterized the concept of imagination throughout the history of philosophy. He instead accomplishes a phenomenological investigation of the *structure* of imagination, analyzing it as an *act* of imagining (a combination of judgements or assumptions, feelings or quasi-feelings) and an *object* imagined, while also suggesting a distinction between a "nominal" act of imagining something and a "propositional" act of imagining *that* a certain state of affairs occurs.

Now, if we still consider these words by Kendall Walton relevant, imagination remains a concept that is still in need of careful investigation:

> What is it to imagine? We have examined a number of dimensions along which imaginings can vary; shouldn't we now spell out what they have in common? – Yes, if we can. But I can't.[32]

In this context, Meinong's refined classification of mental acts allows

- on the one hand, a display of the potentiality of imagination, not only in the appreciation of artworks and arts in general, but in the whole life of human beings, including scientific and moral life;
- on the other hand, a better understanding of the enduring entanglement of perception and imagination, belief and fiction, cognition and emotions in human life. Furthermore, Meinong's account of quasi-feelings anticipates some issues of the contemporary philosophical debate on the so-called paradox of fiction,[33] which might effectively take advantage of a direct confrontation with Meinong's phenomenological work on imagination.

Finally, the cognitive sciences might help us in the near future to identify the analogue within feelings of that "factor of conviction" that differentiates assumptions from judgments, providing us with this still-missing experimental evidence, which will support our account of imaginative feelings in opposition to authentic ones.[34]

References

Brentano, Franz (1874), *Die Psychologie vom empirischen Standpunkt,* Leipzig: Duncker & Humblot.

Deonna, Julien A. / Teroni, Fabrice (2012), *The Emotions: A Philosophical Introduction*, London: Routledge.

Dos Santos, Miguel F. (2017), "Walton's Quasi-Emotions Do Not Go Away", in: *The Journal of Aesthetics and Art Criticism*, 75, 3, p. 265–274.

Eco, Umberto (1995), "Riflessioni teorico-pratiche sulla traduzione", in: Siri Nergaard (ed.), Teorie contemporanee della traduzione, Milano: Strumenti Bompiani, p. 121–146.

Eco, Umberto (2003), *Dire quasi la stessa cosa: esperienze di traduzione*, Milano: Bompiani.

32 Walton 1990, p. 19.

33 Interestingly, both the analytic tradition and the Graz School take David Hume's account of "passions" as a starting point for their reflections on quasi-emotions. See Vendrell Ferran 2010, p. 138-139.

34 Miguel F. Dos Santos argues for a model, compatible with Walton's solution, to empirically test whether quasi-emotions exist. See Dos Santos 2017.

Lauria, Federico (2017), "The Guise of the Ought-to-Be: A Deontic View of the Intentionality of Desire", in: Federico Lauria, Julien A. Deonna (eds.), *The Nature of Desire*, New York: Oxford University Press, p. 139–164.
Massin, Olivier (2017), "Desires, Values, and Norms", in: Federico Lauria, Julien A. Deonna, (eds.), *The Nature of Desire*, New York: Oxford University Press, p. 165–200.
Meinong, Alexius (*GA*), *Alexius Meinong Gesamtausgabe* [*GA*], 7 vols. and supplementary volume, ed. by Rudolf Haller et al., Graz: Akademische Druck- u. Verlagsanstalt 1969–1978.
Meinong, Alexius (1894), *Psychologisch-ethische Untersuchungen zur Wert-Theorie*, in: *GA*, vol. 3, p. 1–244.
Meinong, Alexius (1902), *Über Annahmen*, first edition, partial reprint in: *GA*, vol. 4, p. 385–489.
Meinong, Alexius (1904), *Über Gegenstandstheorie*, in: *GA*, vol. 2, p. 481–530.
Meinong, Alexius (1910), *Über Annahmen*, second edition, Leipzig: J. A. Barth (quoted from reprint in: *GA*, vol. 4); [Engl. transl. *On Assumptions*, ed. by James Heanue, Berkeley / Los Angeles / London: University of California Press, 1983].
Meinong, Alexius (1917), *Über emotionale Präsentation*, in *GA*, vol. 3, p. 285–467; [Engl. transl., *On Emotional Presentation*, ed. by Marie-Luise Schubert-Kalsi, Evanston: Northwestern University Press 1972].
Meinong, Alexius (1923), *Zur Grundlegung der allgemeinen Werttheorie*, in: *GA* vol. 3, p. 471–656.
Mulligan, Kevin (1998), "From Appropriate Emotions to Values", in: *The Monist* 81, 1, p. 161–188.
Raspa, Venanzio (2005), "Forme del più e del meno in Meinong", in: *Rivista di Estetica* 30, 3, p. 185–219.
Reicher, Maria E. (2009), "Value Facts and Value Experiences in Early Phenomenology", in: Beatrice Centi, Wolfgang Huemer (eds.), *Values and Ontology*, Frankfurt: Ontos Verlag, p. 105–135.
Tappolet, Christine (2000), *Émotions et valeurs*, Paris: PUF.
Vendrell Ferran, Ìngrid (2010), "Ästhetische Erfahrung und Quasi-Gefühle", in: Venanzio Raspa (ed.), *The Aesthetics of the Graz School*, Frankfurt: Ontos Verlag, p. 129–168.
Walton, Kendall (1990), *Mimesis as Make-Believe*, Cambridge MA: Harvard University Press.

Gemmo Iocco

Assumptions and Objective Value-Theory: Meinong's Psychology of Valuation

Abstract: The problem of *value* and the related effort in proposing a scientific-objective treatment of it are of the utmost importance in Meinong's philosophical thought. Theoretically speaking, it is common to talk of a conceptual path that led Meinong to abandon the *subjective* perspective he outlined in the early 1890s in favour of an *objective* one developed during the period 1910–1920. Accordingly, the goal of this paper is twofold: on the one hand, we sketch out both the more significant aspects of the subjective value-theory developed in the *Psychologisch-ethische Untersuchungen zur Werth-Theorie* (1894) and the main thesis characterizing the objective value-theory that Meinong approximately started to outline after the publication of the second edition of *Über Annahmen* (1910). On the other hand, we wish to highlight that this change was determined both (1) by his elaboration, from 1899 onwards, of a theory of objects within which the so-called 'dignitatives' and 'desideratives' have a specific ontological relationship, and (2) by the attention Meinong paid to assumptive phenomena.

1 Psychology and Value-Theory: Meinong's Subjective Account

The problem of *value* and the related effort to propose a scientific-objective treatment of it are of the utmost importance in Meinong's philosophical thought. Theoretically speaking it is common[1] to talk of a conceptual path that led Meinong to abandon the *subjective* perspective he outlined in the early 1890s in favour of an *objective* one developed during the period 1910-1920. Meinong's contemplation of value-theory was drawn from Carl Menger's work,[2] although he extrapolates it from Menger's wholly economic nature by focusing on the clearly moral implications possessed by the notion of value. Meinong extrapolated further, at least in

1 In this regard, see Findlay 1963, p. 264–321; Lindenfeld 1980, p. 169–194; Raspa 2013a–b. Instead, Maria Reicher suggests that Meinong already displayed a strong inclination to value-objectivism in his early period (Reicher 2009, p. 114).
2 Carl Menger founded the first Austrian school of value, to which Eugen Böhm-Bawerk and Friedrich von Wieser also belonged (see Smith 1986).

Gemmo Iocco, University of Parma
gemmo.iocco@unipr.it

https://doi.org/10.1515/9783110664850-008

the initial stages, towards the use of a psychological-empirical method proposed by Franz Brentano,[3] who provided an essential conceptual reference for the psychological-philosophical accounts of this kind of problem.

Accordingly, especially in the first stage of his research on value, Meinong pays particular attention to the psychic activities involved in value-experiences; as Howard Eaton[4] demonstrates, the entire problem underlying the value-theory is in fact directed towards a 'general theory of values' as if these were the result of a single psychological process of valuation. Focusing on the concept of valuation amounts to acknowledging that the value of objects is 'felt' and there is the possibility that a valuation does not only concern objects alone but also the experiences through which a valuation occurs. Nevertheless, such an assumption implies, as Wilbur Urban highlights, that "we cannot feel the value of an object without judging or assuming it to have some sort of reality, nor can we pass a normative judgment without at the same time assuming or postulating its conformity with the laws of feeling and will".[5] Thus, the common idea behind this account of value is that value can be understood only by analysing the act through which a subject claims a certain value for objects and experiences.

In his *Psychologisch-ethische Untersuchungen zur Werth-Theorie* (1894), Meinong defines ethics as a special domain in the theory of values, the theoretical fulcrum of which comprises the field of morals. By hinging on the opposition of good and evil, Meinong aimed at analysing a normative meaning of *duty*, in order to define it as 'impersonal'. The core of the general theory of value is not so much an object having specific essential qualities but rather what Meinong defines as 'value-experience', that is to say 'existence-feeling'.[6] According to Meinong, values are not metaphysical constructs: it is for this reason that problems concerning the nature of values are problems that refer to contingent reality. Therefore, the conditions governing the possibility of the existence of values must be traced through the correlation between a subject attributing value to an object and an object in which a certain value is recognised.[7]

---

3 Regarding Brentano's theory of value, see Baumgartner 2002. For an overview of emotion and value in Brentano and Meinong, see Chrudzimski 2009, p. 171–190.

4 Eaton 1930, p. 20.

5 Urban 1908, p. 48.

6 Meinong 1894, p. 26–28.

7 In this regard, Karl Schuhmann states that two apparently incompatible descriptions regarding *value-object* can be identified in the *Investigations* of 1894. According to the first definition, values, are characterized exclusively as objective 'properties' since what does not exist cannot have value; by the other definition presupposing the 'universality of the field of possible value-

Specifically, Meinong articulates how the talk of 'value theory' suggests a very wide research-field within which multiple applications are to be found in the study of economics, aesthetics and morals. Terminologically speaking, to characterize the kind of act through which we apprehend values, Meinong adopts the expression *Werthaltung:*

> Valuation must here mean 'considering something valuable', which it would be a purely intellectual operation to which it would not be appropriate to appeal in the present context since, in order to recognize something as valuable, e.g. as a Value-Object, you already have to know elsewhere to what you actually ascribe the title of 'value' to the corresponding Object. I use the expression 'something valuating' to mean 'holding up' something which nobody wants to identify with 'holding up'. In this sense I can valuate something long before I can conceive the idea of value: valuation is an emotion which, in its independence, is able to giving in itself the natural foundation for that conception.[8]

In other words, Meinong refers to a feeling that guarantees an essential foundation to every ethical valuation. Precisely, the distinctive feature of valuation lies in feeling and not in desire – in contrast to von Ehrenfels's claimed: we desire things because they have a value for us. Their value cannot therefore consist in their being desired. It follows then that it is not through the study of the infinite variety of valued objects that one can throw light on what is hidden behind the expression 'endowed with value', but only by studying the mental attitude that leads one to confer value on things.[9]

From the logical-temporal point of view, *Werthaltung* is prior to desiring [*Begehren*]; further between valuing and desiring there evidently exists a formal difference since it is thanks to *Werthaltung* that the subject is disposed to desire.[10] Again, with reference to the temporal moment, Meinong specifically states that

objects' (*Universalität des Gebietes möglicher Werth-Objecte*), Meiong acknowledges the possibility of an ontological conversion of what exists in what is endowed with value. However, this possibility is shown to be incompatible with the fundamental principles of an *objective* theory of value since it negates the likelihood of an objective ontological differentiation of objects based on the identification of specific structural characteristics. Schumann's account has the merit of demonstrating the interpretive fluctuations present in the first stage of Meinong's reflections and thus confirms the impression that his theory of value of 1894 was only the first stage of a *work in progress* that was to culminate in 1917 with the publication of his essay *Über emotionale Präsentation* (Schuhmann 2001, p. 556).

8 Meinong 1894, p. 27; my translation.

9 Meinong 1894, p. 27–28. As Venanzio Raspa outlines, *valuation* is the psychical fact that is always associated with a value; accordingly, the analysis of value is a psychological analysis (Raspa 2013b, p. 213).

10 Meinong 1894, p. 49.

desire is constitutionally directed towards something that is not present: a healthy person does not recognize any particular value in health, but the moment he or she falls ill, this valuation changes and health becomes of value because it is not present. To be more precise, we are not dealing with a thing that is 'actually'-present, the condition of which is necessary for the desire that refers to the concept of *possibility,* so much as that one desires can be *potentially* verified or can *potentially* exist. Accordingly, value is connected not so much with the *actual valuation* [*aktuelle Werthaltung*] as with the *possible valuation* [*mögliche Werthaltung*]; it is for this reason that in attempting to clarify the relationship between *feeling of value* and *value,* Meinong states that value does not lie in *being evaluated* but in *being able to be evaluated.*[11]

Within desire two distinct forms can be distinguished: 'begehren', understood as a form of desire directed towards something that is still to happen and, in fact, distinct from 'wünschen', an intentional reference that results in the possibility of an orientation towards something that is 'present' or that has 'already happened'. *Wünschen* is, so to speak, a weakened, or less intense, mode of desire. Specifically, what differentiates or distinguishes these two intentional relations is the temporal orientation of the two acts.

Focusing on the internal structure of this emotional form of intentionality, Meinong again takes up and pursues Brentano's classification of mental phenomena in *representation* [*Vorstellung*][12], *judgement* and, emotional *phenomena of love and hate* stating that representation must be understood as a 'psychological presupposition' [*psychologische Voraussetzung*] for emotion. It is not possible to desire something that has not already been represented, and since a representation is always a *representation* of something, it follows that desire must refer to an object.[13] The notion of psychological presupposition brings with it a parallel development leading to the presupposition that it is a logical impossibility to have the first without the second.[14]

11 Meinong 1894, p. 37.

12 In order to distinguish between presentation (*Vorstellung*) and presentation (*Präsentation*), I translate the first one with the term "representation".

13 Meinong 1894, p. 46–47.

14 The impossibility of excluding representation may be the common denominator in the ethical theories of Brentano, Meinong and Husserl, it may be one of the aspects of greater significant differentiation with respect to (for instance) Scheler's attempt to legitimize the a priori foundation of ethics starting from its essential material constitution. Scheler in fact states that in the process of values-perception, representation has no role and this guarantees the autonomy that logic possesses with respect to axiology.

By adopting a more phenomenological lexicon, one can state that the concept of psychological presupposition is nothing other than an alternative formulation of the *foundation law* that regulates the connective relationships between categories of experiences that are themselves formally diverse. That which further characterizes this first account of value is the fact that *value-feelings* [*Wertgefühle*] are always present in relation to the existence or non-existence of something. Within the process that leads to the individualization of value, judgement takes on a role of primary importance because its function is to connect the value-object [*Werth-Object*] with the value-feeling [*Werthgefühl*]. However, it is necessary to pay attention to one specific aspect: the only correct form of judgement that can transform itself into a valuation is that which first passes through a *moment of conviction* [*Überzeugungs-Moment*] and leads thus to *belief*. Furthermore, *belief* is a fundamental element in two ways: first, it is the basis of personal inclination or rejection, and second, it is the foundation of many emotional experiences. Whosoever judges presupposes that "the feeling is right", so what follows is the presupposition that guarantees the theoretical evidence in feelings regarding value, demonstrating that there is a presupposition of the act of judgement and a presupposition regarding the specific content of the judgement. Thus, according to Meinong, emotions directed at value are judicative-feelings [*Urteilsgefühle*].[15]

2 Value-Objectivism and Emotional Presentation

While the main goal of the *Psychologisch-ethische Untersuchungen* of 1894 was to accurately determine the conditions that made 'value-experience' possible, specifically analysing the related mental activities, the development of a theory of objects inevitably demands the adaptation of some conceptual apparatus in order to adequately define characteristics that distinguish value-experience from other experiences also in objective terms.[16]

Structurally speaking, there is an analogy between the general theory of objects and a general theory of values, although from the epistemological point of

15 See Vendrell Ferran 2009, p. 221–224.

16 In this regard Meinong states in the first section of his *Theory of Objects*: "The situation is more complicated in this respect, however, in the case of feelings. There is no doubt that language is somewhat misleading in referring to joy or grief, or to pity or envy, and the like, as being that which one feels. There are also complications in the area of desires, insofar as we think from time that we should revert the possibility of desires which are not desires of anything, despite the linguistic evidence, which is here once again entirely unambiguous" (Meinong 1904, p. 483/ Meinong 1960, p. 76).

view the general theory of objects, understood in that sense as "the doctrine of the sciences" has a greater range as the valued object is only a particular form of the object itself.[17]

Accordingly, the realization of the so-called prejudice *in favour of the actual* [*das Vorurteil zugunsten des Wirklichen*] is the first step in the direction of an effective understanding of that specific category of objects in which one recognizes the quality of values. Stating that "the totality of what exists, including what has existed and will exist, is infinitely small in comparison with the totality of the Objects of knowledge",[18] Meinong displaces – or perhaps, it might be truer to say, eliminates – the boundary line that separates what one can know from what one cannot know, in order to include phenomena such as "non-existence" which in actual reality this does not amount to a strong enough motive for abandoning any attempt of a definition. Undeniably, by this means one can investigate not only a specific object but also the *presuppositions* that define its knowableness. Via the formulation of the *Daseinsfreiheit* – the principle of independence of the *so-being* from the *being* – one does in fact obtain a conceptual, theoretical space for typologies of objects that present themselves in the form of *so-being* [*Sosein*] or in the form of an ontological category that expresses the particular properties possessed by an object. If to exist means having a place within space and time, values, like numbers and states of affairs, do not exist but *subsist* [*bestehen*] since they possess objective qualities that differentiate them from other typologies of objects. Values are objects of a higher order and, to be more precise, are non-independent objects that must have a "lesser" value with which they can be compared: an analogy would be that a relationship must perforce refer to two or more terms.

A particularly interesting aspect when it comes to understanding the innovative nature that Meinong claims for the theory of objects is the relationship established with psychology as the extension of the sphere of knowledge beyond the limits drawn by assumption according to which all *that can be known is in some way or another "real"*. This leads him to ask what role psychology plays in the process of elaborating a *general* theory of objects. Although psychology "is concerned only with real psychological events and not with the merely possible",[19] it has an important clarifying role to play. Indeed, from the *operative* point of view,

17 In this sense, as Findlay states (Findlay 1963, p. 265), the introduction of values in the realm of objects "does not just mean heat and dampness in a cold and dry sphere but also allows the theory of value to gain greater articulation and theoretical depth".

18 Meinong 1904, p. 486 / Meinong 1960, p. 79.

19 Meinong 1904, p. 500.

according to Meinong, "psychology certainly cannot refrain from considering both this capacity to know and also that something outside of reality toward which these characteristic activities are directed".[20] Accordingly, there is not only a *formal* but also a *material* difference between the psychology and theory of objects, since:

> For this reason, it was relevant to note above that the geometrical figure presenting colour-relationships [*Farbenkörper*], as an abstract summary of all of the colours that actually exist in the experience and imagination of men, constitutes one of psychology's concerns. But even this is not strictly true, for such a totality is no more capable of constituting a continuum without the aid of changing events than is a set of points. On the other hand, the idea of colour-relationships conceived in spatial terms [*die Konzeption des Farbnenraumes*] is based *only* on the nature of the Objects in question, and is thus entirely non-psychological, although it doubtless has to do with the theory of Objects. Perhaps the basic difference between the standpoints adopted in the one case and in the other can be traced out immediately in the example, without the assistance of any special comments.[21]

If the distinction between *psychological foundation* and *objective-theoretical* consideration is applied to the value-field, one can claim that a general theory of values presupposes a certain psychological experience, but this presupposition need not be exhaustive in order to understand correctly the experiences themselves.

Indeed, experiences are complex mental phenomena and are distinguished from each other according to the types of objects inherent in them. Exploring the facets of act-quality, experiences can be differentiated as *intellectual* and *emotional*. On the one hand, intellectual experiences are representation [*Vorstellung*] and thought [*Gedanke*]; they form part of the "thought" class of *judgement* [*Urteil*] and *assumption* [*Annahme*]. On the other hand, emotional experiences are divided between *feeling* [*Gefühl*] and *desire* [*Begehrung*]. Each of these forms of experience has a specific object: representations present *objects*, judgements and assumptions present *objectives*, desires present *desideratives*, and feelings *dignitatives*.

Against the backdrop of the distinction between emotional and intellectual experiences, feelings are essentially knowledge-modalities and thus possess an intrinsic cognitive nature, which emerges clearly in relation to the possibility that there is an *emotional presentation* [*emotionale Präsentation*]. The possibility of a feeling-based presentation revolves around a precise definition of the concept of

20 Meinong 1904, p. 495–96 / Meinong 1960, p. 88.
21 Meinong 1904, p. 497 / Meinong 1960, p. 89.

Präsentation, which can be clarified by the following example: if I sense the presence of a red object in my peripheral vision, something in my mind must correspond to that object.[22] However, I am not actively aware of what it is and so for the moment that sensation does not form part of a judgement or an assumption. In this case, it would be more correct to speak of an object that is merely presented to us from the current moment, in one sense, as passive: this relationship Meinong defines as 'präsentieren'. What is presented is never adequate to our understanding because the idea cannot apprehend the object by itself but must assume its object via another experience. Through emotional presentation, an *affectively descriptive* content is presented to the subject in a way that it seems analogous to that operates in the intellectual experience. The distinction is not so much *qualitative* as relative to progression, since the capacity of feelings to act as means of knowledge is less than that of which presentations are capable. For the sake of simplicity, if we compare two statements such as: 'the temperature is nice' and 'the temperature has gone up', the first expression has a cognitive value that is less than that of the second, although the first statement gives us a form of understanding.[23] By the same token, it follows that *an object has value* not because a subject attributes value to it, but rather because it *is worthy of the interest* that the subject gives it: value therefore is what is presented thanks to our experience of value.

Moreover, intellectual and emotional experiences have both a *presuppositional-relationship* [*Voraussetzungsverhältnis*]: if, for example, someone likes a colour or a melody, his or her emotion is connected to an object and therefore to its content, but as Meinong claims, this content is an *integral part* of the representation of the colour or melody but not in the same way as the content of an emotion can be considered the content of a representation.

> Between the presuppositional objects and the experience with whose presupposition we are dealing, there is, we may admit, *another* objective relationship in virtue of this presuppositional relationship [*Voraussetzungsverhältnis*]. When the smell of a flower pleases me, when I am rejoiced by the German and Austrian victories of 1915, and when, at the same time, my wishes are directed to the return of peace among civilized nations, this certainly does not merely mean that certain representations and thoughts, to which certain objects correspond, have been made the presuppositions of my feelings or desires. It means also that, in virtue of these presuppositions, certain feelings and desires are directed to certain objects and that these objects may accordingly be called the objects of the feelings and de-

22 On Meinong's account of self-presentation, see Lindenfeld 1980, p. 182–83; Marek 2003, p. 163–73.
23 Meinong 1912, p. 278.

> sires in question. Whenever objects (that is, objecta or objectives) are emotionally approached, objects are thus connected with the approaching experiences. These probably were connections considered when objectivity [*Gegenständlichkeit*] was attributed, not only to representations, but to all other experience, and was classed as a characteristic common to everything psychic.[24]

Generally speaking, psychological presupposition indicates the dependence relationship between specific classes of object, or more specifically in the case of the perception of values, the emotional presentation is configured as a dependency of subsistence [*Unselbständigkeit des Bestandes*] since values do not exist but *subsist*.

Accordingly, in distinguishing the psychological presupposition of intellectual experiences from the psychological presupposition of emotional experiences, Meinong claims for feeling a specific epistemological function. Further, this consideration demonstrates that the type of presentation that operates in the affective sphere is a *partial presentation* [*Partialpräsentation*] since one does not deal with a total experience but with an element of the same, which in a certain way helps to direct experience towards something that is neither experience itself nor an experience in its entirety.

> If, therefore, there are proper contents of feelings and desires, then the preconditions for a partial presentation which is a content-presentation are also given in the case of emotional experiences. That these are more than mere possibilities is shown by some everyday attributions, as when people talk about a refreshing bath, fresh air, oppressive heat, disturbing noise, beautiful colour, funny or sad, boring or entertaining, stories, sublime works of art, valuable people, good resolutions, etc. There is no question as to the close relationship of such attributes to our feelings. And there is no question that these attributes are fully analogous to other proprieties which are familiarly presented by representations.[25]

When one formulates judgements such as 'the sky is blue' or 'the sky is lovely', according to Meinong, in both cases we attribute a *property* to the sky, but while in the first case the understanding of the 'blue' property is understood via representation, in the second the understanding of 'lovely' comes via a feeling: it is the feeling that thus reveals itself as a factor that determines the presentation.[26] It follows that, 'in favourable circumstances', feeling may function to present the contents of objects: feelings are *vehicles* of understanding, even if, from the

24 Meinong 1917, p. 315 / Meinong 1972, p. 27; slightly modified translation.

25 Meinong 1917, p. 317–18 / Meinong 1972, p. 28; slightly modified translation.

26 Meinong 1917, p. 317.

qualitative point of view, they do not reach the apodictic levels that distinguish thoughts.

As far as the object of presentation is concerned, if one judges it necessary to have an object of judgement, when for instance, one celebrates the victories of a certain country on the battlefield, one has an emotional experience that is possible only to the extent that one considers the victory of that country on the battlefield to be a value; indeed,

> it is easily seen that a feeling or desire is never said to be justified or unjustified *per se*, but always relative to an object to which the emotion in question is directed, and which is its presuppositional object.[27]

In trying to define with the greatest accuracy the declination in the emotional sphere of the distinction between *act* and *content*, Meinong shows that, contrary to what the expression "representative action" might appear to mean, what one is actually dealing with is not so much an activity as *passive* experiences: presentations and emotions are passive operations whilst judgements and desires are active operations.

> The word "act" is, above all, not meant to indicate activity. The contrast between active and passive, which seems to me, despite objection, from various sides to be of fundamental importance, holds in fact only within the confines of what is entitled to be called an "act". Representations especially, whose act-aspect we are now discussing, are passive experiences, like feelings; whereas judgments and assumptions are active experiences.[28]

Such a distinction, relative to the type of process involved, leads one inevitably to argue that objects present themselves as *imbued with emotional significance.*

With the aim of determining those objects that are presented through *emotionale Präsentation,* one must remember that while the subsistence and *extra-being* [*Aussersein*] of objects is known a priori, it is the object of a direct apprehension [*direktes Erfassen*] or of a 'quasi-experience' through which the theory of objects finds itself in a position directly analogous to that of empirical science as it follows a path *from below upwards.*[29] The methodological validity of this inductive process, Meinong adds, is largely evident if one takes into account aesthetic objects which, being objectified through an emotional presentation, possess a formal configuration that can be more easily described. If one analyses the prop-

27 Meinong 1917, p. 414 / Meinong 1972, p. 114.
28 Meinong 1917, p. 340–41 / Meinong 1972, p. 50; slightly modified translation.
29 Meinong 1917, p. 387.

erties of 'lovely', one cannot do otherwise than reveal that the analysis itself, besides reflecting something of the properties of 'lovely' (as long as it can be attributed to an object), has need of further characteristics, for example whether it is endowed with a certain proportion; beauty is thus grasped by *a perceiving-like intuition* [*wahrnehmungsartige Anschaulichkeit*].[30] If this possibility is extended to the entire theory of values, emotional presentation is configured as a modality of intuition that allows one to recognize objects that have an emotional nature; this intuition is a *sui generis* experience, which has important affinities with intellectual experiences, and further, allows for intuition in dealing with objects that are constituted differently.

Moreover, emotional presentation is a 'partial' presentation because it does not correspond to a complete understanding such as occurs in the case of intellectual experiences; thus, knowledge coincides with *penetrative apprehension of something that is factual or possible* and is structured like an internally justified judgement [*innerlich berechtigtes Urteilen*] in fact:

> Knowledge is the penetrative apprehension of something factual or possible, wherein this apprehension has that moment of internal legitimacy which is known as evidence.[31]

At stake, there is the evidence of an emotional sphere, albeit this evidence is not to be found in emotional experiences themselves but rather in judgements, since dignitatives and desideratives cannot be completely apprehended by their presentatives but require a supplementation by assumptions or judgements.[32] Indeed, if correct judgements are evident, correct emotions are *legitimate*. However, the margin that exists between "legitimacy" [*Rechtmässigkeit*] and "correctness", with reference to their respective objects, leads Meinong to state that "the legitimacy of the feeling is no more constituted thereby than the sum of the angles in a triangle is constituted by the insight that it is equal to two right angles. "[33]

> On a closer look, it is easily seen that a feeling or desire is never said to be justified or unjustified per se, but always relative to an object to which the emotion in question is directed, and which is its presuppositional object. If we have a feeling, for example one of pleasure, no one will say that it is justified or unjustified to feel such a pleasure, but one may well be justified in being pleased with *A* and unjustified in being pleased with *B*. This runs parallel to the fact that no one will think it correct or incorrect to represent the object "light" per se, but that he will think it correct to represent aluminium as "light" and incorrect to represent

30 Meinong 1917, p. 388.
31 Meinong 1917, p. 403 / Meinong 1972, p. 106.
32 Meinong 1917, p. 406.
33 Meinong 1917, p. 412 / Meinong 1972, p. 113.

> lead as "light". If such representation is said to be correct since the judgment whose predicate is apprehended by the idea is true, it is quite appropriate to call a feeling justified if it presents an object whose designation is the predicate of a true judgment and whose presuppositional object's name is the subject. I am justified [right, *im Rechte*] in being pleased by the successes of the allied Central Powers during the present war to the extent that such a feeling of pleasure presents an object that can rightly be attributed to the successes in question, since it in fact applies to them. It can therefore be said in general: if *P* is an object presented by an emotion *p*, then it is justifiable to attach the emotion *p* to an object *A* if *P* in fact applies to *A*, and the judgment "*A* is *P*" is therefore correct. "Correct" and "incorrect", when they are in this sense attributed to emotions, doubtless do not mean the same as "correct" and "incorrect" when used in connection with judgments, yet the new use of the expressions derives from their use in connection with judgments. An evidence-analogue for feelings and desires is therefore not required.[34]

Epistemological access to the world of values is possible only by indirect paths since values constitute themselves as objects that are not perceived as *subsistent*. Values are related to knowledge *a priori* but at the same time they are configured in such a way as to encourage access from below [*von unten*], perhaps because, as Meinong states, the empirical path has an intrinsic directionality or finality *a priori*. Accordingly, the entire science of values must take its first steps in the empirical field and stay as close as possible to that path: in this way, the general theory of values may be constituted as something more than a simple theory of objects. The fundamental epistemological question is thus directed to understanding how one can recognize something that is empirically induced as being "a priori". In fact, according to Meinong, all judgements elaborating material objectivity previously given "a priori" status are due to the relationships between the content and its object that necessarily correspond to each other.

In the emotional sphere concerning the concept of impersonal value it is not sufficient to focus on the concept of foundation [*Fundierung*]. Clearly, if in the intellectual presentations of the concepts of "red" and "green" can be found the basis of the object of "difference" [*Verschiedenheit*], in emotional presentations such a basis is not logically possible because one would be obliged to claim that all emotional presentations present *impersonal* values.

> Red and green together provide the foundation for the object Difference (*Verschiedenheit*); comparison, however, gives rise to the production of that representation which presents the object Difference (this time intellectually), and by this presentation we apprehend the fact of the foundation of one object on others (*Fundierungstatbestand*). Here, too, foundation and presentation are completely different things. They are, however, naturally, even if amazingly, connected by the fact that the knowledge-experience,

34 Meinong 1917, p. 414–15 / Meinong 1972, p. 114–15; slightly modified translation.

> rendered possible by the presentation, apprehends with evidence the subsistent relation of foundation. This leads us to ask whether, in the case of emotional presentation, the connection could not also be made by similar means. The situation is not in all respects similar. If emotional presentation carried in itself evidence for such foundation, each value-experience would reveal an impersonal value exactly as each set of comparisons, made under sufficiently favourable circumstances, yields a completely impersonal similarity or difference. But we have already noted that there is a possibility or justified surmises. In the light of such surmises, it is no longer puzzling that, out of sufficiently congruous presentations, a knowledge of such foundations, or of such impersonal values, can be attained.[35]

Emotional presentation is configured as a means for cognition [*Erkennungsmittel*] only if one presupposes that: a) what is presented through an experience of value must not be attributed to objects as if this is their specific property, b) objects already possess that type of property that allows them to be the efficient cause of the same presentation.[36]

From the axiological point of view, Meinong's objective value account has the merit of drawing attention to certain aspects of fundamental importance that allow us to proceed from *simple description of experiences* through which values are assumed to the legitimisation of their normative claims. Or, to be exact, a) the value of an object does not exclusively depend on its specific characteristics since the value of an object very often depends on the value of another; b) the value that a specific object has is never precisely determined but is proportionate to its *usefulness* – the law of marginal utility [*Grenznutzen*]; c) all things being equal, it is the subject who shows the greater interest that will decide the value of an object. These conditions clarify the purely relativistic nature of the laws that a structured constitutive process will be capable of determining. However, this does not present any obstacle to stating that such relationships possess specific aspects of objectivity. Indeed, if we limit ourselves to the definition of value according to which value is the *ability of an object to attract the interest of the subject* we run the risk of mistakenly believing that such an ability depends on the rightness or wrongness of the judgements that represent the psychological presumption of the constitution of the same value. The additional features that an object must possess to be defined as "lovely" lie outside the fact of its existence and may then be constituted as *objective* or else as an object of a higher order:

> The objects "pleasant", "true", and "good" also lack self-sufficiency and depend upon the presuppositional objects of the feeling which present them. They likewise do not

35 Meinong 1917, p. 457 / Meinong 1972, p. 151; slightly modified translation.
36 Meinong 1917, p. 432 ff.

> require a plurality of substrates. The same can be said of the exclusive objects of desire, that is, oughtness and end [*Sollen* and *Zweck*]. This affords a modest beginning to the characterization of the objects of emotional presentation, but it is, nonetheless, a genuine first step when we say: The objects of emotional partial presentation are objects of higher order.[37]

The passage from a subjective theory of value to an objective one is precisely grounded in the possibility of an impersonal account of value within which, on the one hand, value is presented to the subject, and on the other it obtains an independent ontological status.

3 The Role of Assumptions in Meinong's Value-Theory

Amongst the diverse functions carried out by assumptions[38] there is one that is of primary importance in the consideration of value-experience. This role is evident in such cases in where an assumption allows the presentation of an objective in those forms of activity that are of an emotional nature, or else when assumptions provide the objective for feelings, desires and valuations. Further, it is evident that the theory of assumptions has decisive theoretical implications for the development of the objective value-experience account proposed by Meinong, starting from the publication of the second edition of *Über Annahmen*.

Indeed, in Chapter IX of *Über Annahmen* (1910), entitled '*Zur Begehrungs- und Wertpsychologie*', Meinong reaffirms that it is misleading to define desire as the point of origin for value-experiences from the moment that what one desires has not yet been realised. To be precise, before one desires something, one experiences a sort of "projection" through which the subject represents him or herself *as it would be* [*wie es wäre wenn*] before the situation he or she is considering is realised in fact. The specific state of this *pre-valuation* experience is none other than an assumption. An assumption, therefore, is a sort of *motivational process* [*Motivationsvorgang*] that precedes desire. According to Meinong's subdivision of experiences into intellectual and emotional, as within intellectual experiences

37 Meinong 1917, p. 392 / Meinong 1972, p. 95–96.

38 Assumptions are experiences of an intellectual nature and provide an intermediary epistemological link between representation and judgement: regarding simple representation, assumption comprises an experience aimed at choosing between positive and negative. However, this does not in itself provide the evidence that characterizes judgement. Assumptions have their specific object or objective [*Objektiv*] which is not a simple objective but a *state of affairs* (for a concise overview of Meinong's theory of assumption see Jacquette 2015, 41–57 and Donato-Rodriguez 2016).

assumption takes up an intermediate position between representation and judgment, in emotional experiences assumption takes up an intermediate position between representation and feeling.

> In the attempt at an initial characterization of the state of affairs regarding assumptions, we found ourselves more than once compelled to describe assumptions as mental facts that occupy a kind of intermediate position between representations and judgments. I now believe, I must assert, in a quite analogous way, that there are also mental facts that must be assigned a kind of mid-position between representations and feelings.[39]

Specifically, the typologies of experience that come into play here are defined by Meinong as *fantasy feelings* [*Phantasiegefühle*] and act as psychological presuppositions for so-called fantasy desires [*Phantasiebegehrungen*].[40] The fact that fantasy-feelings and fantasy-assumptions are experiences of fundamental importance in the emotional sphere rests on their ability to constitute themselves in the form of *as if* [*als-ob*]. Take, for example, a person who, aiming to keep his or her riches safe, decides to bury them: it follows that in the moment he / she becomes aware that his or her riches have been discovered and removed, the possibility of being robbed, which was previously a remote representative likelihood, becomes an extremely *vivid* one. This cognitive augmentation is determined by a fantasy feeling that possesses, as its foundation, an assumption of belief.

Accordingly, assumptions guarantee to the emotional theory of value a specific depth that leads one to realise that feelings of value present themselves not only as 'judgement-feelings' but occasionally also as 'assumption-feelings'.[41] In a narrow sense, Meinong suggests the expression 'quasi-feeling' [*Quasi-Gefühl*] to indicate those specific forms of imaginative feelings that demonstrate both a fundamental cognitive function and a decisive *motivational* power within the value-desire connection. In addition, in this case, Meinong highlights a certain similarity between the emotional and intellectual spheres considering that

> the contrast of feelings-in-earnest and imaginative feelings governs no less than the whole domain of feelings, in the same way the contrast of judgments and assumptions governs the whole domain of those intellectual experiences that stand over against objectives as their means of apprehension.[42]

---

39 Meinong 1910, p. 309 / Meinong 1982, p. 221.

40 As far as Meinong the relationship between fantasy and emotion is concerned, see Raspa 2013b and Vendrell Ferran 2013.

41 Meinong 1910, p. 333.

42 Meinong 1910, p. 337 / Meinong 1982, p. 240.

The decisive role played by assumptions in the emotional-field is stressed also in *Über emotionale Präsentation*, where, in order to characterize the concepts of feeling and desire, Meinong identifies two forms of formally different understanding. Indeed he distinguishes a *contemplative understanding* [*ein kontemplatives Erfassen*], which occurs through representation and assumption from a *penetrative understanding* which instead occurs through judgement. Applying this subdivision to the emotional sphere, one realizes that assumptions can act as presuppositions for feelings of imaginative value. Assumptions thus reveal themselves to be *surrogates* for "serious" value feelings [*Werternstgefühle*] constituting themselves as true *aids* to the contemplative understanding of values. But, although assumptions are factors that help us to determine the value intensity [*Wertgrösse*], on its own, value can be recognised only through an operation of a "penetrative" nature aimed at the epistemological fulcrum of that same act.

Further, while feelings of value and aesthetical feelings are both a sub-species of the "*Inhaltsgefühle*" type, they are not differentiated as presupposed objects, but rather as different types of *Gefühlspräsentation*. In this regard, Meinong states that it is not possible to doubt, for example, that a person who is interested in the original nature of a document for important legal reasons has an attitude that is markedly different from a person who is interested in its originality for historical reasons. In the former case, one can talk of *Wertgefühl* and in the latter of *Wissensgefühl*. Regarding feelings of the cognitive type the contrast between affirmation and negation required in judging these presuppositions, has no role to play other than that which is necessary to feelings of value. Between these two schemes, there is the same relationship subsisting between the factual [*tatsächlich*] and the possible which at its *ontic* level [*Seinshöhe*] refers to the content of a judgement. Value-judgements are hence acts of valuation having an existing-content or an assumed-content, that is an objective: in the latter case, we speak of judgement-like assumptions. These experiences' modalities ensure to the valuing subject an 'as-if experience' that is highly advantageous to make a choice.

In conclusion, it seems clear that whilst, during the 1890's Meinong stated that the fundamental experience of value was the *Werthaltung* – that is the feeling produced by a belief in the existence or non-existence of the evaluated object –, by the study of assumption, the genesis of the feeling of value can be traced in *Wertung* or in that form of valuation that can find its foundation not only in a judgement of existence but also in an assumption.[43] The introduction of the con-

43 In this regard, Meinong writes: "Three expressions enter into primary consideration, 'valuing' [*Wertung*], 'valuation' [*Werthalten*], and 'evaluation' [*Bewerten*]. In my earlier publications, I have

cept of assumption should not only be seen as an enlargement and thus a deepening of the theory of knowledge proposed by Meinong, but as a conceptual improvement that proves to be of primary importance also in describing the fundamental components that come into play in the emotional sphere. Correspondingly, in the development of an objective account of value-experience, the key role of assumption resides in the fact that it enables Meinong to consider value-experience not so much as a 'mediated' experience, grounded in representation, but as a direct modality of value-apprehension that, however, requires assumptions or judgements to outline itself as a complete experience.

References

Baumgartner, Wilhelm (2002), "Franz Brentano: The Foundation of Value Theory and Ethics", in: J. Drummond and L. Embree (eds.), *Phenomenological Approaches to Moral Philosophy*, New York: Springer, p. 119–138.

---

already attempted to define the latter two by suggestions as to their use. Expressly, and on more than one occasion, I have characterized valuation [*Werthalten*] as the feeling elicited by the conviction of the existence or non-existence of an objectum. More in passing, I have characterized evaluation [*Bewerten*] as the merely intellectual apprehending of a value, in short, as value-judgment. Yet, the expression 'valuing' [*werten*] has for some accidental reason remained peculiarly foreign to my personal linguistic usage. As a consequence, it has remained open for some new use, and I now gladly accept it as the only term available for doing justice to the terminological needs created by the above statements, at least in s way that is not ungrammatical- so 'valuing'[*Wertung*] is what I want to call the attitude of someone who reacts to an assumption of the existence or non-existence of an objectum with the imaginative feeling repeatedly mentioned above. Valuing [*Wertung*] is the counterpart to valuation [*Werthalten*], then; it is distinguished from the latter by the fact that judgment essential to valuing is replaced by an assumption and by the fact that the judgment-feeling essential to valuing is replaced by an assumption and by the fact that the judgment-feeling essential to valuing is replaced by an assumptive quasi-feeling, i.e., by an imaginative feeling. As mentioned, I do not delude myself that this arrangement is free of all arbitrariness. But if one is to try, in the interests of theory, to differentiate between three expressions that are so closely connected with each other pretheoretically, then it seems to me that in no case can things be brought off with some element of arbitrariness or convention. Moreover, it seems a reasonable hope that those who have been accustomed to employ 'valuing' and valuation indifferently, as has usually been the case until now, can more easily assent to my suggestion according as they find our description in terms of an assumption and an imaginative feeling appropriate to the facts that they have hitherto not given any separate name" (Meinong 1910, p. 333–34 / Meinong 1982, p. 238). On this issue, see also Centi 2009, p. 83–9. In order both to distinguish between Valuation and Valuing and to be consistent with the terminology adopted in the text, I reversed the Heanue's translation of the two terms.

Brentano, Franz (1969), *Vom Ursprung sittlicher Erkenntnis*, Hamburg: Meiner.
Centi, Beatrice (2009), "Relations, Quasi-Assumptions and Material Aprioris", in: B. Centi and W. Huemer (eds.), *Values and Ontology. Problems and Perspectives*, Frankfurt / Paris / Lancaster / New Brunswick: Ontos Verlag, p. 45–105.
Chrudzimski, Arkadiusz (2009), "Brentano, Marty and Meinong on Emotions and Values", in: B. Centi and W. Huemer (eds.), *Values and Ontology. Problems and Perspectives*, Frankfurt / Paris / Lancaster / New Brunswick: Ontos Verlag, p. 171–190.
Donato-Rodriguez, Xavier (2016), "Meinong's Theory of Assumptions and its Relevance for Scientific Contexts", in: M. Antonelli and M. David (eds.), *Existence, Fiction, Assumption. Meinongian Themes and the History of Austrian Philosophy, Meinong Studies / Meinong Studien* 6, p. 141–174.
Eaton, Howard O. (1930), *The Austrian Philosophy of Value*, Norman: University of Oklahoma Press.
Findlay, John N. (1963), *Meinong's Theory of Objects and Values*, Oxford: Clarendon Press.
Jacquette, Dale (2015), *Alexius Meinong, The Shepherd of Non-Being*, Cham: Springer.
Lindenfeld, David F. (1980), *The Transformation of Positivism. Alexius Meinong and European Thought 1890–1920*, Berkeley: University of California Press.
Marek, Johann C. (2003), "On Self-Presentation", in: C. Kanzian, J. Quitterer, E. Runggaldier (eds.), *Persons. An Interdisciplinary Approach. Proceedings of the 25th International Wittgenstein Symposium*, Wien: öbv & hpt, p. 163–173.
Meinong, Alexius (1894), *Psychologisch-ethische Untersuchungen zur Werth-Theorie*, in: *Alexius Meinong Gesamtausgabe*, Band III: *Abhandlungen zur Werttheorie*, hrsg. v. R. Haller und R. Kindinger, Graz: Akademische Druck- u. Verlagsanstalt, 1968, p. 3–244.
Meinong, Alexius (1904), *Über Gegenstandstheorie*, in: *Alexius Meinong Gesamtausgabe*, Band II: *Abhandlungen zur Erkenntnistheorie und Gegenstandstheorie*, Graz: Akademische Druck- u. Verlagsanstalt 1971.
Meinong, Alexius (1910), *Über Annahmen*, Leipzig: J. A. Barth, in: *Alexius Meinong Gesamtausgabe*, Band IV: *Über Annahmen*, Graz: Akademische Druck- u. Verlagsanstalt 1969.
Meinong, Alexius (1912), "Für die Psychologie und gegen den Psychologismus in der allgemeinen Werttheorie", in: *Logos* 3. Reprinted in: *Alexius Meinong Gesamtausgabe*, Band III: *Abhandlungen zur Werththeorie*, hrsg. v. R. Haller und R. Kindinger, Graz: Akademische Druck- u. Verlagsanstalt, p. 267–282.
Meinong, Alexius (1917), "Über emotionale Präsentation", in: *Sitzungsberichte der Kaiserlichen Akademie der Wissenschaften in Wien*. Philosophisch-historische Klasse, Band 183, 1917, Abhandlung II. Reprinted in: *Alexius Meinong Gesamtausgabe*, Band III: *Abhandlungen zur Werththeorie*, hrsg. v. R. Haller und R. Kindinger, Graz: Akademische Druck- u. Verlagsanstalt, p. 283–467.
Meinong, Alexius (1960): "The Theory of Objects", Engl. transl. by I. Levi, D. B. Terrell, and R. M. Chisholm, in: Chisholm, R. M. (ed.), *Realism and the Background of Phenomenology*, Glencoe (Ill.): Free Press, p. 76–117.
Meinong, Alexius (1972), *On Emotional Presentation*, translated, with an introduction by M.-L. Schubert Kalsi, with a foreword by J. N. Findlay, Evanston (Ill.): Northwestern University Press.
Meinong, Alexius (1983), *On Assumptions*, edited and translated, with an introduction by J. Heanue, Berkeley / Los Angeles / London: University of California Press.
Menger, Carl (1871), *Grundsätze der Volkswirtschaftslehre*, Wien: Wilhelm Braumüller.

Raspa, Venanzio (2013a). "Valori personali e valori impersonali", in: S. Colloca (ed.), *The Value of Truth / The Truth of Value*, Milano: LED, p. 125–143.
Raspa, Venanzio (2013b), "Meinong on Aesthetic Objects and the Knowledge-Value of Emotions", in: *Humana.Mente Journal of Philosophical Studies* 25, p. 211–234.
Reicher, Maria E. (2009), "Value Facts and Value Experience in Early Phenomenology", in: B. Centi and W. Huemer (eds.), *Values and Ontology. Problems and Perspectives*, Paris / Lancaster / New Brunswick: Ontos Verlag, p. 105–136.
Schumann, Karl (2001). "Value Theory in Ehrenfels and Meinong", in: L. Albertazzi, D. Jacquette, R. Poli (eds.), *The School of Alexius Meinong*, Aldershot: Ashgate, p. 541–570.
Smith, Barry (1986), "The Theory of Value of Christian von Ehrenfels", in: R. Fabian (Hrsg.), *Christian von Ehrenfels. Leben und Werk*, Amsterdam: Rodopi, p. 150–171.
Urban, Wilbur (1908), "What is the function of a general Theory of Value?", in: *The Philosophical Review* 17, 1, p. 42–62.
Vendrell Ferran, Íngrid (2009), "Meinongs Philosophie der Gefühle und ihr Einfluss auf die Grazer Schule", in: A. Schramm (ed.), *Meinong Studies / Meinong Studien* 3, p. 199–240.
Vendrell Ferran, Íngrid (2013), "Ästhetische Erfahrung und Quasi-Gefühle", in: V. Raspa (ed.), *The Aesthetics of the Graz School, Meinong Studies / Meinong Studien* 4, p. 129–168.

Venanzio Raspa

Meinong on Emotional Truth

Abstract: Truth is a semantic notion. However, it can involve our whole being, including our emotional part. This article explores the somewhat puzzling notion of emotional truth from a Meinongian perspective. According to Meinong, knowledge is justified true judgment, but a kind of justification belongs to emotions, too. A value is objective when its valuation does not rest on false premises (judgments) as in the case of superstition. In a similar way, an emotion which is triggered by a judgment is justified if the related judgment is justified as well. Emotional truth is the kind of truth which gives rise to a justified knowledge-feeling, where a knowledge-feeling is that feeling by which we feel the value of a truth, that is, of a justified judgment which in turn justifies the emotion connected to it.

1 Why is it Important to Talk about Emotional Truth?

Truth is a semantic notion. However, I will try to understand it not only from a semantic point of view, but also as something that involves our whole being, both our intellectual and emotional faculties. Finally, it should become clear that the concept of emotional truth has a political import.

This notion has recently been investigated by Ronald de Sousa, who published in 2011 a book precisely entitled *Emotional Truth*. In this book, he presents in full extent an idea which he has developed in previous articles; in vain, however, one will search in these works for a reference to Meinong. Yet, there are some general analogies between Meinong's and de Sousa's thought, including the following theses: emotions allow us to know values and are constitutive of the latter; values are both objective and relative; or emotions can be appropriate or inappropriate in relation to a given object.[1] This last point has also been suggested by Kevin Mulligan, according to whom emotions can be said to justify axiologyical beliefs;[2] on the other side, emotions themselves are justified by

1 Cf. de Sousa 2011, p. 20 ff., 33ff., 103.
2 Cf. Mulligan 1998.

Venanzio Raspa, Università degli Studi di Urbino Carlo Bo
venanzio.raspa@uniurb.it

https://doi.org/10.1515/9783110664850-009

perceptions and beliefs, and are said to be appropriate if, and only if, the axiological judgments they support are correct.[3]

I confine myself to these few remarks. Rather than making a comparison between Meinong and de Sousa, or understanding Meinong – as suggested by Mulligan – in a line of thinking which appeals to appropriate emotions, a line of thinking which starts with Brentano and Meinong and arrives at contemporary analytic philosophy,[4] my purpose is narrower and is limited to Meinong's concept of emotional truth, which has not yet been investigated. Moreover, while de Sousa is interested in showing that we can speak of true or false emotions, and that, therefore, truth and falsity do not exclusively belong to judgments and propositions, Meinong's focus is rather on characterizing the truth, or at least *some* truths, as emotionally coloured.[5]

In two recent papers, I have made a passing reference to the notion of emotional truth at the conclusion.[6] In the present writing, on the basis of previous researches on Meinong's theory of values and emotions, I will investigate more in detail what has been outlined there. Meinong does not write much about emotional truth, neither a book chapter nor a paper, but only a note in his personal copy of *Über emotionale Präsentation* [*On Emotional Presentation*] (1917). I will show that, starting from this note, it is possible to draw a more complete picture of what Meinong means by emotional truth. I will use the method according to which, just as starting from a single object (from an individual) we may discover a whole world, or a portion of the world, in the same way it should be possible to start from a single concept and identify a network of concepts, by following the thread of relations which connects the concepts or entities in question. Let us read the passage in question:

> It seems that "true" cannot naturally be coordinated with "beautiful" and "good" [...]. It is said, indeed, that a flower is beautiful, that an action is good, that a judgment is true. But the flower is beautiful because it is blue and has such and such a shape. The action is good because another person's interest is put before one's own. But in that case "true" seems to be a parallel case to "blue" or "altruistic." It does not designate the proper object [*Eigengegenstand*] presented by the knowledge-feeling [*Wissensgefühl*] but at most the object borrowed [*den angeeigneten Gegenstand*] by this feeling. In reality we are here only dealing with matters of expression. "True" has, in the first place, an intellectual meaning to which

3 Mulligan's concept of appropriateness is very close to that of justification I use here; actually, Mulligan (1998, p. 162) himself recognizes such an identification.

4 A part of this story is offered in Mulligan 2017.

5 Meinong (1894, *GA* III, p. 88) explicitly states that true and false do not apply to feelings.

6 Cf. Raspa 2016, p. 208–209; 2018, p. 45.

> correspond the notion of truth as pertaining to objectives and the notion of truth as pertaining to experiences [*der gemäß es den Objektivbegriff und Erlebnisbegriff der Wahrheit gibt*]. But, in addition to this intellectual meaning, there is an emotional meaning [*emotionale Bedeutung*] of "truth" in virtue of which something is called "true" if it attracts a justified knowledge-feeling [*berechtigtes Wissensgefühl*]. In this sense we can speak of a truth-feeling [*Wahrheitsgefühl*] which can genuinely be coordinated with a feeling of beauty and with (so to speak) a goodness-feeling.[7]

In order to understand what Meinong means by emotional truth, we have to understand this passage, a very dense one, which is rich in specific terms. What follows is an explanation of it. The result will be a network of concepts that together constitute, if not a theory, at least the sketch of a theory. A reference to the notion of emotional truth is also to be found in the "Selbstdarstellung [Self-Portrait]" (1921), one of Meinong's last writings.[8] This gives evidence that the concept belongs to his late reflection; however, the theoretical reasons that led him to speak about emotional truth can be found in earlier writings.

The following three theses are fundamental for the development of such a notion. (1) The object is independent of the subject, but it is accessible only by means of the subject, more precisely by means of mental experiences (representations, thoughts, feelings, and desires). (2) The value is related to an emotion: if a thing has a value for a subject, it does not leave him indifferent, the subject gives a valuation of it, which is tantamount to saying that the object is capable of arousing emotions in the subject. (3) Emotions possess not only an evaluative character, but also a cognitive one. The first thesis is a general one, which characterizes the whole Meinongian philosophy; the other two should become clear in the course of what follows.

According to Aristotle, an emotion is a mental and physical state which is intentional and complex. It consists of cognitive-evaluative elements and can also affect cognitive and evaluative processes.[9] Meinong shares this view. In addition, he argues that emotions allow us to know what the world is like. In his view, certain features of the objects, which he interprets as specific ideal objects of higher order, can be apprehended only by means of emotions; without the active role of the latter, they would remain unknown to us. This means that, in Meinong's opinion, there are specific forms and characteristics of the objects, which we apprehend through our emotions, but this fact does not make them

7 Meinong 1968, *GA* III, p. 750 [1972, p. 169]; the English translation has been slightly modified (V.R.).

8 Cf. Meinong 1921, *GA* VII, p. 42.

9 I explain this definition in Raspa 2016, p. 177–181.

subjective. And if emotions have a cognitive function, then they have a relation to truth. Speaking of emotional truth amounts to connecting the emotional element with the cognitive one.

2 Truth in a Semantic Sense and Grounded Knowledge

Truth is not produced by the subject, but is known by him. According to Meinong, truth is grounded knowledge. In this sense, it is expressed by means of an internally true judgment, and is therefore the result of an intellectual operation. Neither things and their properties nor experiences and their determinations are true (or false), except judgments and, in very special circumstances, assumptions, but judgements are said to be true on the basis of their objectives.[10]

Meinong calls 'objectives' the specific objects of judgments and assumptions.[11] He does not provide a definition, but rather a description of the objectives, bringing out the distinctive features that, as a whole, distinguish them from other types of objects. Objectives are ideal objects of higher order, that at most subsist, and are truth-bearers like the states of affairs. They are higher-order objects because they require objecta (i.e., objects of representations) as their presuppositions just as – as we will see – judgments require representations as their constituent parts.

A true objective is a subsisting objective; it designates a fact and hence it is a factual (*tatsächlicher*) objective; on the opposite side, a false objective is non-factual (*untatsächlich*).[12] In *Über Annahmen* [*On Assumptions*] (1910), Meinong does not give a definition of factuality (*Tatsächlichkeit*),[13] but merely a description of it. Factuality is a modal property of objectives, the limit of a line of different magnitudes corresponding to different degrees of possibility.[14] From an epistemological point of view, a factual objective is an objective apprehended with evidence and certainty;[15] as there are degrees of certainty, there is also a gradual scale (*graduelle Abstufbarkeit*) for evidence.[16] Still in *Über Möglichkeit und Wahrscheinlichkeit* [*On Possibility and Probability*] (1915) Meinong does not define factuality, but illustrates it with examples: that there are the antipodes, that the magnetic

---

10 Cf. Meinong 1915, *GA* VI, p. 38–40.
11 Cf. Meinong 1904, *GA* II, p. 387 [1960, p. 80]; 1910, *GA* IV, p. 44 [1983, p. 38]; 1915, *GA* VI, p. 26–27.
12 Cf. Meinong 1910, *GA* IV, p. 69 [1983, p. 55].
13 Cf. Meinong 1910, *GA* IV, p. 70–71 [1983, p. 56].
14 Cf. Meinong 1910, *GA* IV, p. 89 [1983, pp. 68–69].
15 Cf. Meinong 1910, *GA* IV, p. 85 [1983, p. 65].
16 Cf. Meinong 1910, *GA* IV, p. 91 [1983, p. 69].

needle is deflected by the galvanic current, are factual objectives; while that there are spirits, that the round square exists, that radioactivity and magnetism are the same thing, are non-factual objectives. Between these two extremes, there are objectives whose factuality is open, which Meinong calls 'subfactual' (*untertatsächlich*) and are merely possible. Possibility is a modal property of objectives.

> Factuality is the maximum of possibility, and possibility is "factuality of lower degree", decreased or still incomplete factuality.[17]

The line of possibility can also be referred to as the line of factuality.[18]

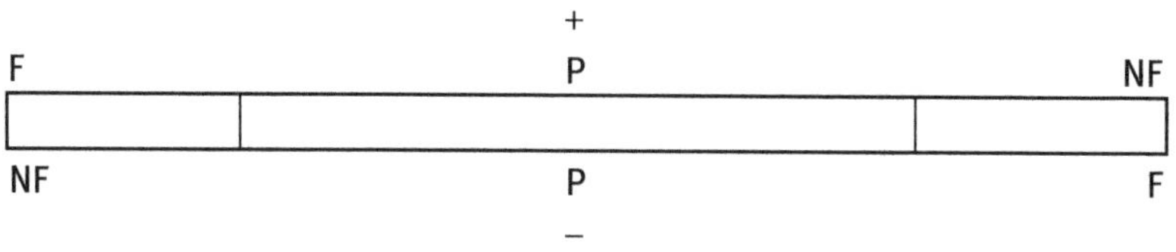

Truth is thus a property of objectives.[19] An objective is true if it is factual, but truth is not the same as factuality, because truth requires apprehension, that is, a cognitive act.[20] Since a factual objective is expressed or apprehended by a judgment or an assumption, truth is not independent from the fact that it is known. "Knowledge is internally justified judging (as opposed to merely externally justified judging) that satisfies the demands of truth or probability",[21] because truth is the upper limit of probability.[22] Factuality is an ontological notion, truth is an epistemological one.

In the *Viertes Kolleg über Erkenntnistheorie* [*Fourth Course of Lectures on Theory of Knowledge*] (1917/18), Meinong reasserts that there is no knowledge without truth,[23] that "knowing is true judging [*Erkennen ist wahres Urteilen*]",[24] and that a true judgment is a judgment which apprehends a factual objective. This does not mean that each true judgment is knowledge, because a judgment can accidentally be true and is, therefore, irrelevant to knowledge. There is a feature

17 Meinong 1915, *GA* VI, p. 92.
18 Cf. Meinong 1915, *GA* VI, p. 95.
19 Cf. Meinong 1915, *GA* VI, p. 40.
20 Cf. Meinong 1917/18, *GA. Ergänzungsband*, p. 346.
21 Meinong 1917, *GA* III, p. 404 [1972, p. 106]. Cf. also Meinong 1915, *GA* VI, p. 416.
22 On this see Raspa 2005, p. 205ff.
23 Cf. Meinong 1917/18, *GA. Ergänzungsband*, p. 340.
24 Meinong 1917/18, *GA. Ergänzungsband*, p. 343.

which distinguishes knowledge from simple truth: knowledge is the true judgment which is internally true (*innerlich wahr*). Meinong means that apprehension is grounded knowledge, that is, justified true judgment. On the contrary, something can be true and not be known as such. In this case, it is a mere extrinsic truth.[25]

To sum up, there are two notions of truth as well as two related notions of knowledge: there is a notion of truth (or of knowledge) as pertaining to objectives (*Objektivgbegriff der Wahrheit*) and a notion of truth (or of knowledge) as pertaining to experiences (*Erlebnisbegriff der Wahrheit*).[26]

3 Truth as Value, the Cognitive Function of Emotions and their Justification

Let us consider our opening passage again. We can read the first proposition like a question: Is it possible to coordinate 'true' with 'beautiful' and 'good'? At the end of the passage Meinong writes: "a truth-feeling can genuinely be coordinated with a feeling of beauty and with a goodness-feeling." This suggests that he understands 'true' like 'beautiful' and 'good,' that is, as a value, which as such attracts, or is capable of attracting, some feelings. And this is the definition of emotional truth: "something is called 'true' if it attracts a justified knowledge-feeling." We have then to understand these new notions of knowledge-feeling and of truth as value. We begin with a short exposition of Meinong's concept of value.[27]

Truth as value. In *Psychologisch-ethische Untersuchungen zur Wert-Theorie* [*Psychological-Ethical Investigations on Value Theory*] (1894), Meinong maintains that the value an object has lies in the possibility of its being evaluated (*Werthgehalten-werden-können*). Thus an object has a value insofar as it is able to arouse in a subject the factual grounds (*die tatsächliche Grundlage*) for a value-feeling (i.e., pleasure or displeasure for the existence or the non-existence of something).[28] One year later, in "Über Werthaltung und Wert [On Valuation and Value]" (1895), Meinong states again that "the value of an object can be [...] de-

25 Cf. Meinong 1917/18, *GA. Ergänzungsband*, p. 344–345.
26 Cf. Meinong 1915, *GA* VI, p. 414–415; 1917, *GA* III, p. 419 [1972, p. 118]; 1917/18, *GA. Ergänzungsband*, p. 346.
27 For more details on the content of this paragraph, see Raspa 2013a, of which I use some parts in the following pages.
28 Cf. Meinong 1894, *GA* III, p. 37.

fined as its capacity to be appreciated by an intellectually and emotionally normal subject."[29] Valuation (*Werthaltung*) is that psychical fact (a feeling) that is always associated with a value. If something is valuable to me, then I will be related to it in such a peculiar way that the thing will acquire a special meaning for me.[30] It does not leave me indifferent, but arouses an emotion.

In Meinong's view, the object has certain properties that, once they are acknowledged by the subject, allow for its valuation; the object possesses these properties independently of being evaluated. To put it differently, value is a second-order property, or, as Meinong will say in *Über emotionale Präsentation*, a higher-order property.[31] If we apply this definition of value to a property like beauty, it follows that an object is beautiful if it is able to arouse in a subject the grounds for a positive aesthetic feeling. It does that because it possesses some other properties. Like the property 'red,' the property 'beautiful' requires a *substratum*, something of which it is a property; but it requires in addition another property or set of properties as its basis. The property 'beautiful' is then non-independent from its basis and dependent upon the characteristics of such a basis. This is the meaning of the statement "the flower is beautiful because it is blue and has such and such a shape." As a property, blue is an object of higher order; as an aesthetic property, beautiful is an object (so to speak) of second order. The same holds for truth as value: it is able to arouse in a subject the grounds for a positive truth-feeling, which is a kind of value-feeling;[32] moreover, it is non-independent from its basis and dependent upon the characteristics of such a basis.

The cognitive function of emotions. Emotions perform an intellectual function by means of which specific properties like beautiful or good – which require a valuation – can be apprehended. Otherwise – i.e., without taking emotions into account – the apprehension of such properties would be beyond reach.[33]

Now, it often happens that we attribute value to something which has no value or, conversely, we refrain from attributing value to something which has it. A subject can attribute to a twig the value of a divining rod. In this case, by means of valuation, a value is ascribed to the object which depends only on the subject, being thus totally subjective. When valuation is concretely applied to an object (as for example to a medicinal plant), then its value is objective.[34] Value therefore

29 Meinong 1895, *GA* III, p. 248.
30 Cf. Meinong 1894, *GA* III, p. 26.
31 Cf. Meinong 1917, *GA* III, p. 392, 394 [1972, p. 96, 97–98]. See also Meinong 1921, *GA* VII, p. 22 [1974, p. 228–229].
32 Cf. Meinong 1894, *GA* III, p. 50.
33 Meinong's point of view about emotions and values has been developed in Tappolet 2000.
34 Cf. Meinong 1894, *GA* III, p. 78–81.

is objective when its valuation does not rest on false premises (judgments) as in the case of superstition. This meaning of 'objective' shows a strong analogy with the meaning of 'justified' (*berechtigt*) we have seen above and to which we will return soon.

In the cognizing process, representations have a 'basic' position. According to the Brentanian intentionality thesis, which Meinong endorses, any thought requires an object that is thought of, any feeling an object that causes pleasure or displeasure, any desire an object towards (or against) whose being or non-being a subject is directed. Meinong calls this object a "presuppositional object" (*Voraussetzungsgegenstand*), or a "borrowed object" (*angeeigneter Gegenstand*), which may but does not have to be apprehended by means of a representation. However, it must be apprehended by means of a psychical experience which – when not itself a representation – presupposes a representation.[35] A judgment is a non-independent experience, which – in order to exist – is in need of another experience that will function as its "psychological presupposition" (*psychologische Voraussetzung*).[36] For Meinong as for Brentano, judgment always requires a representation, while the converse does not hold. Moreover, judgment has a double object: the one *about* which we judge (i.e., the representational object) and the one *that* is what is judged (i.e., the objective).[37] The former is the presuppositional or borrowed object, the latter is the proper object (*Eigengegenstand*). It is impossible to judge without judging *about* something (the representational object), but it is equally impossible to judge without judging *that* something is the case (the objective). This implies that a judgment cannot directly apprehend its object and that it requires another experience, towards which it is non-independent. This prerequisite experience presents[38] the object which is judged about, so that in the simplest case it is a representation, which works as a psychological presupposition.[39] In this sense, any experience 'is based' on a representation.

Emotions, together with the object towards which they are directed, make up a complex: if someone smells a flower's scent, then there is a representation, which is directed to a certain object and which serves as the emotion's psychological presupposition. By means of this presupposition, the emotion is directed towards an object, which is hence the object of that emotion, and not merely the

35 Cf. Meinong 1917, *GA* III, p. 294 [1972, p. 8–9].
36 Cf. Meinong 1917, *GA* III, p. 290 [1972, p. 6].
37 Cf. Meinong 1910, *GA* IV, p. 43–44 [1983, p. 38].
38 About the act of presenting, see fn. 45.
39 Cf. Meinong 1917, *GA* III, p. 351–352 [1972, p. 60–61].

object of a representation (or of a judgment). The object is connected with the psychical experience towards which it is directed.[40]

The proper object of an emotion is not that to which it is addressed, that is, the borrowed object. According to Meinong, emotions are characterised by a double object, a presuppositional (or borrowed) object and a proper object. When I get carried away by the sadness of a melody, it holds true that sadness is a feeling, but not the sadness of a melody, because only sentient beings can feel, not objects like melodies; the sadness of a melody is an object of higher order; in this case, it is the proper object which is attributed to the presuppositional object, that is, the melody, to which our emotion is addressed. The latter is also constituted by the experiences of judgment and representation, which allow me to apprehend the melody, but not the sadness, which can instead be apprehended only by means of a feeling.[41]

Failure to recognize this duplicity between borrowed and proper object was often the cause of errors and misunderstandings. It was so, when it led to identifying the proper object of judgments with the object of representations, until the objective was recognised as the proper object of judgments which is absolutely distinct from the borrowed object (the objectum) on which it is founded. And it was cause of error as it led to identifying the object of emotions with that from which the emotion is aroused or to which it is directed. But things stand differently.

Justified emotions. We have seen that emotional experiences are means for knowing objects. But knowing, Meinong writes, is always an intellectual operation. An emotional experience cannot alone apprehend an object. It can do so only if it is connected with an intellectual experience as its psychological presupposition.[42] The property 'true' belongs to the objective and, by extension, to the judgment. Now, if knowledge is justified true judgment, and if – although under certain conditions – emotions are means of knowing, one may wonder whether they, too, possess the "moment of justification" (*Berechtigungsmoment*). A positive answer would be tantamount to saying that emotions have a relation to truth.

However, we know that emotions need representations and judgments as their psychological presuppositions. Does this mean that they are not at all connected to truth? What can their justification consist of? Meinong's solution is the following: the justification of emotions should be sought in non-emotional experiences, that is, in judgments. It will be a mediated justification. Of course, the

40 Cf. Meinong 1917, *GA* III, p. 314–315 [1972, p. 26–27].

41 Cf. Meinong 1917, *GA* III, p. 324, 365–366 [1972, p. 35, 72–73].

42 Cf. Meinong 1917, *GA* III, p. 403 [1972, p. 106].

moment of justification should not be sought in obvious cases such as gratitude or benevolence, which guarantee their value by themselves, but in the many other cases where the valuation is mediated.[43] In order to understand Meinong's argument, we should bear in mind that emotional knowledge is a complex fact: it always presupposes both representations (presupposed by all experiences) and judgments (which may be true or false) – if the apprehension should be complete.[44] And here, presented in a condensed form, is Meinong's argument.

Traditionally and correctly, it has been denied that ideas can be true or false; nonetheless, it is sometimes said that someone has a right or false idea, which means that this idea is likely to give rise to a true or false judgment. If an emotion takes the place of a representation as the means of presentation (*Präsentationsmittel*),[45] then, when the corresponding judgment is justified or not, the justification can be attributed to the emotion as well. An emotion is then never justified or unjustified *per se*, but it is so in relation to an object towards which it is directed, which is its presuppositional object. Nobody would say that it is justified or unjustified to feel joy, but one may be justified or unjustified in rejoicing in something or in a certain fact – one is unjustified in rejoicing in the pain of raped children. Meinong synthesises this idea as follows:

> If *P* is an object presented by an emotion *p*, then it is justifiable to attach the emotion *p* to an object *A* if *P* in fact applies to *A* (*dem A zukommt*), and the judgment "*A* is *P*" is therefore correct. "Correct" and "incorrect", insofar as these are said of emotions in this sense, doubtless do not mean the same as "correct" and "incorrect" in the case of judgments, but are nevertheless taken over from the latter.[46]

What has been said is easily demonstrated in aesthetic terms, but it is also valid from an epistemological perspective. An emotion is justified if the judgment

43 Cf. Meinong 1917, *GA* III, p. 408 [1972, p. 109–110].
44 According to Meinong, apprehension (*Erfassen*) can be complete (*fertig*) or incomplete (*unfertig*). The latter belongs to representations, which are the psychical experiences indispensable for any apprehension but lack the active character requisite for a complete apprehension. This active character consists – as Aristotle already argued – in saying something about something. Therefore, a complete apprehension belongs only to judgments and assumptions which express a thought. Cf. Meinong 1917, *GA* III, p. 287–289, 402–403 [1972, p. 3–5, 105–106].
45 'Presentation' (*Präsentation*) is a technical term, meaning the act of a psychical experience (for ex. a representation) that offers an object to thought. According to Meinong, not only representations and judgments but also emotions can play this role. For more details on the theory of presentation, which I have intentionally avoided exposing here, see Raspa 2013a, p. 216ff.; 2016, p. 186ff.
46 Meinong 1917, *GA* III, p. 414–415 [1972, p. 115]; the English translation has been modified (V.R.).

which attributes the proper object of the emotion (a predicate like sadness) to its presuppositional object (for example, a subject like a melody) is justified. To take Meinong's own example:

> I am right to rejoice in the successes of the allied Central Powers during the present war to the extent that the feeling of joy presents an object that can rightly be attributed in a judgment [*zugeurteilt werden*] to the successes in question, since it in fact applies to them.[47]

This kind of mediated justification, to which emotions can aspire, allows us – without resorting to any notion of evidence – to assign an objective character to emotions within the cognitive process, and to determine when an emotion is not justified – like, for instance, when an Italian entrepreneur has taken pleasure from the earthquake that struck the city of L'Aquila on 6th April 2009. Kevin Mulligan generalises Meinong's idea as follows:

> If indignation about the fact that *p* is correct, it is correct because it is unjust that *p*.[48]

According to Meinong, "whoever believes something also believes that he is right in his belief",[49] and therefore, someone who connects a certain value-feeling to a certain presuppositional object, will condemn the opposite behaviour.

4 Emotional Truth

Emotions connected to truth attach to true objectives and judgments. However, it must not be forgotten, even if it can be disregarded here, that truth is the limit of probability and that this also attracts similar feelings, namely, truth-feelings, which have true objectives as their presuppositional object.[50] Without denying the subjectivity of feelings, Meinong says that, when emotional knowledge is justified, the known object (the dignitative or value) is not relative to the subject, is not subjective or 'personal' (in his terminology), but is objective, absolute or 'impersonal'.[51] Meinong denies that falsehood could play a role as presuppositional object. He believes that ideals like love, justice, or truthfulness, are unanimously recognised as values. A similar recognition is to be found in the epistemological domain, if one accepts the notion of truth as value and, consequently, that of

47 Meinong 1917, *GA* III, p. 414 [1972, p. 115]; the English translation has been modified (V.R.).
48 Mulligan 2017, p. 499.
49 Meinong 1917, *GA* III, p. 421 [1972, p. 120].
50 Cf. Meinong 1917, *GA* III, p. 417–418 [1972, p. 117–118].
51 On Meinong's view about personal and impersonal values, see Raspa 2013b.

emotional truth: in such a case, the connection between feeling and presuppositional object is knowable a priori. That is why it may be said, in the popular way of speaking, that truth is "felt".[52]

Beside a notion of truth as pertaining to objectives and a notion of truth as pertaining to experiences, there is therefore a value-notion of truth (*Wertbegriff der Wahrheit*). If value is capable of attracting feelings, the same holds for truth. As we have seen, Meinong's classification of feelings encompasses not only sensuous, aesthetic, and value-feelings, but also knowledge-feelings, which, having to do with knowledge, have judgments as their psychological presuppositions. Their respective objectives are connected to feelings which may arise, for example, whenever a long-hidden truth is discovered. I think for example of the many mysteries of the recent history of Italy (the fascist massacres of Piazza Fontana and Bologna, the Ustica disaster, the murders of Ilaria Alpi and Miran Hrovatin, of Giulio Regeni, and Stefano Cucchi). The knowledge-feeling is that feeling by which we *feel* the value of truth, that is, of a justified judgment which in turn justifies the emotion connected to it. Here the notion of truth as value or the notion of emotional truth, that is, something which attracts a justified knowledge-feeling, assumes significance.

From this idea we could proceed to develop an aesthetic of concepts. A theory establishes itself if it speaks not only to the thought, but also to the heart, if we know of it, along with what is true, also what is good and beautiful. We apprehend a true value, one for which one is willing to get involved, and possibly to die,[53] not only with our intellect, but also through our emotions. We discover that a truth is relevant for us by means of the emotion connected to it. There are truths that are not relevant, like that which says that the pen in my hand is longer than eight centimetres.

The success of a theory depends not only on the truth it encompasses, because sometimes the theory that establishes itself is simply false; it also depends on the fact that the truth, or the proposition that one believes to be true, involves the subjects emotionally. This will concur to explain the phenomenon that young people are attracted by irrational and nefarious theories like neo-Nazism and racism, or choices like those of foreign fighters, as well as the phenomenon of whole peoples who did embrace a given political concept.

---

52 Cf. Meinong 1917, *GA* III, p. 418 [1972, p. 118].

53 An example given by Meinong (1917, *GA* III, p. 408 [1972, p. 110]), which recalls the sad war times in which he lived as he wrote and published *Über emotionale Präsentation*, concerns "those who continue fighting and dying on a sinking skip, instead of surrendering themselves".

Meinong distinguishes between being true and being taken to be true, a distinction which is important for us, and indicates a direction in which we can further investigate. He says that a value is objective when its valuation does not rest on false premises (judgments) as in the case of superstition; that knowledge is justified true judgment, and the emotion which is attracted from a judgment is justified if the connected judgment is justified as well.

> If *A* is the presuppositional object of a value-feeling *p* which presents the proper object *P* (*Eigengegenstand*), then the simultaneous givenness (*Zusammengegebensein*) of the objects *A* and *P* gives reason to presume that *A* has *P*.[54]

This means that we are justified in making surmises starting from our actual valuations and experiences, as well as in proceeding from weak surmises to increasingly stronger ones on the basis of our experiences and, possibly, of other indirect moments added for our purposes.

On this basis, an aesthetic of concepts is a task that lies before us, it is a work that shows how some theoretically justified truths involve us emotionally and are therefore also beautiful; on the contrary, other judgments are unjustified both theoretically and regarding the emotions they arouse and which involve the whole being of an individual. However, two other cases are also possible: one can hold a falsehood as justified or, conversely, a truth as unjustified, according to the context and other factors which colour it emotionally. Such errors can persist and have a long duration if they are connected to emotions that act as reinforcement. A theory, a belief, a philosophy, an ideology, a religion ceases to have appeal, if it loses its emotional impact.

References

De Sousa, Ronald (2011), *Emotional Truth*, Oxford: Oxford University Press.

Meinong, Alexius (1894), *Psychologisch-ethische Untersuchungen zur Wert-Theorie. Festschrift der K. K. Karl-Franzens-Universität zur Jahresfeier am 15. November 1894*, Graz: Leuschner & Lubensky; repr. in: *GA* III, p. 1–244.

Meinong, Alexius (1895), "Über Werthaltung und Wert", in: *Archiv für systematische Philosophie* I, p. 327–346; repr. in *GA* III, p. 245–266.

Meinong, Alexius (1904) "Über Gegenstandstheorie", in: *Untersuchungen zur Gegenstandstheorie und Psychologie*, hrsg. von A. Meinong, Leipzig: J. A. Barth, p. 1–50; repr. in *GA* II, p. 481–530 [Engl. transl.: Meinong (1960)].

54 Meinong 1917, *GA* III, p. 422 [1972, p. 121].

Meinong, Alexius (1910²), *Über Annahmen*, 2., umgearbeitete Auflage, Leipzig: J. A. Barth; repr. in *GA* IV, p. 1–389, 517–535 [Engl. transl.: Meinong (1983)].
Meinong, Alexius (1915), *Über Möglichkeit und Wahrscheinlichkeit. Beiträge zur Gegenstandstheorie und Erkenntnistheorie*, Leipzig: J. A. Barth.; repr. in: *GA* VI, p. xv–xxii, 1–728, 777–808.
Meinong, Alexius (1917), *Über emotionale Präsentation*, in: *Kaiserliche Akademie der Wissenschaften in Wien*. Phil.-hist. Klasse, Sitzungsberichte, CLXXXIII, 2. Abh.; repr. in: *GA* III, p. 283–476 [Engl. transl.: Meinong (1972)].
Meinong, Alexius ([1917/18]/1978), *Viertes Kolleg über Erkenntnistheorie*, in: *GA, Ergänzungsband: Kolleghefte und Fragmente. Schriften aus dem Nachlass*, p. 337–401.
Meinong, Alexius (1921), "A. Meinong [Selbstdarstellung]", in: *Die deutsche Philosophie der Gegenwart in Selbstdarstellungen*, hrsg. von R. Schmidt, Bd. 1, Leipzig: Meiner, p. 91–150; repr. in *GA* VII, p. 1–62 [Partial Engl. transl.: Meinong (1974)].
Meinong, Alexius (1968–1978), *Alexius Meinong Gesamtausgabe*, hrsg. von R. Haller und R. Kindinger gemeinsam mit R. M. Chisholm, Graz: Akademische Druck- und Verlagsanstalt.
Meinong, Alexius (1960), "The Theory of Objects", translated by I. Levi, D. B. Terrell, and R. M. Chisholm, in: *Realism and the Background of Phenomenology*, ed. by R. M. Chisholm, Glencoe (Ill.): Free Press, p. 76–117.
Meinong, Alexius (1972), *On Emotional Presentation*, translated, with an introduction by M.-L. Schubert Kalsi, with a foreword by J. N. Findlay, Evanston (Ill.): Northwestern University Press.
Meinong, Alexius (1974), "Meinong's Ontology", "Meinong's Life and Work", in: R. Grossmann, *Meinong*, London / Boston: Routledge & Kegan Paul, p. 224–229, 230–236.
Meinong, Alexius (1983), *On Assumptions*, edited and translated, with an introduction by J. Heanue, Berkeley / Los Angeles / London: University of California Press.
Mulligan, Kevin (1998), "From Appropriate Emotions to Values", in: *The Monist*, 81, 1, p. 161–188.
Mulligan, Kevin (2017), "Incorrect Emotions in Ancient, Austrian & Contemporary Philosophy", in: *Revue philosophique de la France et de l'étranger*, 142, 4, p. 491–512.
Raspa, Venanzio (2005), "Forme del più e del meno in Meinong", in: *Il pregiudizio a favore del reale. La teoria dell'oggetto di Alexius Meinong fra ontologia e epistemologia*, ed. by C. Barbero and V. Raspa, special issue of *Rivista di estetica* XLV, n.s. 30, 3, p. 185–219.
Raspa, Venanzio (2013a), "Meinong on Aesthetic Objects and the Knowledge-Value of Emotions", in: *Meinong Strikes Again. Return to Impossible Objects 100 Years Later*, ed. by L. Mari and M. Paolini Paoletti, special issue of *Humana.Mente. Journal of Philosophical Studies*, 25, p. 211–234.
Raspa, Venanzio (2013b), "Valori personali e valori impersonali. Sulla teoria del valore di Alexius Meinong", in: *The Value of Truth / The Truth of Value*, ed. by S. Colloca, Milano: LED, p. 125–143.
Raspa, Venanzio (2016), "Col corpo capisco. Emozioni e conoscenza in Aristotele e Meinong", in: *Tra corpo e mente. Questioni di confine*, a cura di B. Centi, Firenze: Le Lettere, p. 175–209.
Raspa, Venanzio (2018), "Le vicende del contenuto attraverso Bolzano, Twardowski e Meinong", in: *Paradigmi*, XXXVI, N. S., No. 1, p. 31–48. DOI: 10.30460/89365.
Tappolet, Christine (2000), *Émotions et valeurs*, Paris: Presses Universitaires de France.

Vendrell Ferran, Íngrid (2009), "Meinongs Philosophie der Gefühle und ihr Einfluss auf die Grazer Schule", in: *Meinong Studies / Meinong Studien*, vol. 3, Frankfurt *et al.*: Ontos Verlag, p. 199–239.

Erwin Tegtmeier

Meinong's Realist Analysis of Perception

Abstract: Early Phenomenology (Brentano, Meinong) may be understood as an attempt to overcome representationalism and idealism and to promote realism. While Brentano sees no way to defend a realist analysis of outer perception, Meinong eventually developed such an analysis. He vindicates outer perception by introducing a second kind of evidence (evidence for presumption) and by assuming in line with psychophysics that relations of comparison and other higher-order objects can be transferred from mental phenomena to physical noumena. This transference is also justified by the apriori knowability of higher-order objects. Meinong needs a transference from phenomena to noumena because he takes Galileo's argument from physics for granted, to the effect that bodies do not have the perceptual qualities they seem to have. However, that does not make such a big difference because the exact determination of the qualities of physical objects follows the same principles as those Meinong adopts from psychophysics. In this article it is argued that the concept of presumptive evidence can be integrated into epistemological realism.

1 Towards Epistemological Realism

Brentano, Meinong's teacher, holds that outer perception, in contrast to inner perception, is not evident and is therefore cognitively worthless.[1] Brentano assumes a quality of mental acts he calls "evidence". Being evident is to have that quality which amounts to indubitability. In contrast to Brentano, the later Meinong, in his 1906 book entitled *Die Erfahrungsgrundlagen unseres Wissens*, tries to get outer perception some evidence (in Brentano's sense). Meinong even conceives another kind of evidence for that. In this article I try to understand that development of Meinong's ontology of perception as an attempt to get closer to a tenable epistemological realism. Grossmann diagnoses that Meinong wanted to overcome his earlier Kantianism and had all the theoretical means to succeed if only he would have seen that the argument of physics (as Grossmann calls it) is not sound.[2]

1 Brentano 1970, 2nd part, section II.
2 Grossmann 1974, chapter VII.

Erwin Tegtmeier, Universität Mannheim
tegtmei@rumms.uni-mannheim.de

https://doi.org/10.1515/9783110664850-010

Admittedly, the phrase "epistemological realism" does not play a conspicuous role in Meinong. He takes the realism issue to be about the existence of an external world[3] disregarding whether we know that external world or not. Thus, even Kant would pass as a realist since he assumes an external albeit unknowable "thing-in-itself".

Meinong as well as Brentano are epistemological realists in the first place insofar as they overcome Descartes' representationalism as well as the various forms of subjective and objective idealism which were dominating 18th- and 19th-century philosophy. Realism is the antithesis to all three of them. What particularly comes to light in Meinong is the view that our cognition is on the whole correct while idealism implies that our cognition is fundamentally wrong (most notably, what appears to be physical is in fact mental, according to idealism) and representationalism implies that we cannot know anything outside the mind since we can never compare the representation of something with the thing represented. Note that epistemological realism is defined in terms of correctness, not in terms of correspondence. Correspondence belongs to representationalism, not to epistemological realism.[4] One can be more specific and distinguish between local and global correctness. A single misperception is a case of local incorrectness. If there are no colours, all perceptions are wrong and we face global incorrectness. Local incorrectness can be removed, global incorrectness is insurmountable. Thus understood, realism claims that our knowledge is globally correct.

2 From Classical Representationalism to Idealism

Roughly speaking, the problems that led to representationalism and, finally, to idealism may be summarised as follows:

The new 17th-century particle physics, in addition to long-standing skeptical arguments against outer perception, gave rise to the view that perceptual objects don't have most of the properties we perceive at them and that these properties are merely reactions of our minds or our sensory apparatus to micro-physical structures. For the representationalists, the contents of our minds (for which they adopted Plato's term "ideas") are pivotal to our knowledge of the world, although their role and status is merely that of representations. The massive sceptical or even agnostic consequences of this epistemology brought idealism into the

3 Meinong 1906, p. 80.
4 See Tegtmeier 2014.

arena. Such an idealism already was to be found in French and British Empiricism, according to which the material objects given to us in perception are mere complexes of ideas. While in Empiricism there sometimes was an ambiguity as to whether complexes of ideas merely stand for the material objects they represent or are identical to them, this ambiguity disappeared in Berkeley and Kant.

Kant thinks that he, contrary to Berkeley, escapes idealism by assuming "an unknown and unknowable cause of our sensations". Unknowables, however, are irrelevant to epistemology. Kant pretends to fully vindicate perception but his attempt to do so is not to be taken at face value, and his own declaration that he wanted to limit reason in order to make room for religious faith is telling enough. How can outer perception be locally correct if its objects are complexes of mental contents?

The Aristotelian view of perception had been realistic, explaining that in perception the essence of the perceived object is transferred into the perceiver. It was Descartes who created a gap between mind and material objects. He held that mind knows directly only what is in it (Principle of Immanence), while material objects are known only indirectly, by means of ideas in the mind. Thus he assumed a rather weak connection between the perceivers and the perceived material objects. Kant went so far as to claim that we can know only the ideas in our minds and that the objects we perceive are merely ideas in the mind produced by the latter in accordance with its capacities and needs. Descartes deliberately led philosophy to scepticism. Kant pretended to save us from scepticism. At any rate, he initiated the opposite of realism, namely the very influential German Idealism.

3 Overcoming Classical Representationalism and Idealism

Brentano significantly departed from classical representationalism. Although he rejected Descartes' representationalism, he nevertheless adopted Descartes' principle, according to which only what is evident (indubitable) should be considered an instance of knowledge in the proper sense of the term. It is from this perspective that he rejected Kant's transcendental way of thinking as being a "philosophy of prejudices" (*Philosophie der Vorurteile*). Brentano started his attempt to overcome idealism by sharpening the distinction between the mental and the physical, which had been blurred by the French and British Empiricists. Already they, as was mentioned, and not only Kant had analysed material objects as complexes of ideas and thus, strictly speaking, as mental entities.

Brentano introduced intentionality as a criterion of the mental early on, but he took the decisive step towards epistemological realism when he later assumed intentionality to be a specific relation which holds directly between a mental act and its object. Moreover, Brentano's principle that the object of a mental act is never part of that act is directed against idealism, since it rules out any mental activity processing mental contents such as Kant's so-called "synthesis of the understanding". Importantly, Brentano's mental acts are no actions or intellectual operations.

Brentano developed an ontology of relations with special reference to intentionality. But since in his view the objects of some mental acts do not exist, he concludes that the intentional relation is best seen as an attribute of mental acts only. Brentano's ultimate view, which extends his analysis of intentionality to all kinds of relations, was that relations belong only to one of the two relata. However, a one-sided intentional relation, in which the object only serves to modify the respective attribute of the mental act, does not seem to allow us to grasp the object by means of that mental act. Meinong's intentional relation appears more appropriate to support a realist claim to the effect that we grasp the object of a mental act. The reason it is so is that Meinong's intentional relation is two sided, building upon the mental act and the object, and holds necessarily and internally.[5]

The vindication of perception supports also epistemological realism's claim that *all* knowledge is globally correct. The title of Meinong's book, *Perceptual Foundations of Knowledge* (in English), emphasises the epistemologically crucial role of perception. Meinong's realism is revealed by the following points: 1. he explicitly rejects the principle of immanence and defends the belief in an external world; 2. he tries to vindicate outer perception; 3. he especially advocates the claim that perception has really noumenal objects; 4. he claims that relations of comparison between phenomenal objects are transferable to noumenal objects. I will return to 3 and 4 later.

4 Epistemological Realism and Evidence

At first, it was difficult for me to connect Brentano's concept of evidence to realism, especially since Bergmann and Grossmann argued that evidence was a representationalist notion and that Meinong's second kind of evidence was misconceived and superfluous.[6] But finally I saw a way to do it, and this, I believe, gave me a better understanding of what epistemological realism is about.

5 Meinong 1902, p. 126f.
6 Bergmann 1967, p. 422f.; Grossmann 1974, p. 134f.

Meinong himself establishes a connection between evidence and realism insofar as he argues that, without evidence, all our knowledge is threatened and could fail, more precisely could be false, since without evidence we do not know whether an "objective" (*Objektiv*), to which an act of judgement is intentionally related, really obtains. Thus Meinong thinks that a judgement need not only be true but also known to be true. Otherwise, its correctness is not secured. However, not all knowings can be immediately known to be true. Most are usually verified by relating them logically to other knowings. Without immediately evident knowings such verifications would be circular, Meinong argues.

As was made clear above, evidence is used here in an absolute sense, not in the relative sense of evidence for some assumption or conclusion. A judgement has evidence just in case it is evident (in Brentano's sense) and does not need to derive its justification from other judgements. Thus there is a connection between the absolute and relative sense of evidence. What has evidence can serve as evidence for something else. Evidence in Brentano's sense could be defined as incontrovertible correctness. As we shall see, it is pivotal to Meinong's epistemology.

Meinong modifies the concept of evidence he inherits from Brentano considerably and he disagrees with Brentano as regards which kinds of mental acts have evidence. Brentano denies outer perception any evidence while Meinong tries to secure some evidence for it. He introduces an additional kind of evidence, and this allows him to somehow advocate a realist view of outer perception – i. e., to show that the latter has some kind of correctness. As far as evidence is concerned, Brentano refers us exclusively to inner perception. He thus completely rejects realism as regards outer perception.

As was mentioned already, Meinong distinguishes between evidence for certainty and evidence for presumption (*Vermutungsevidenz*). The former corresponds roughly to Brentano's concept of evidence. Brentano wanted to base all empirical sciences on inner perception, but Meinong makes clear how limited the evidence of inner perception is. Only the inner perception of the nearly simultaneous is evident with certainty, since introspective knowing which goes beyond the present mental state has to rely on memory. However, memory does not have evidence with certainty but only evidence for presumption. Meinong adds certain judgements to those having evidence for certainty, judgements to which Brentano does not attribute such evidence – namely, judgements of comparison based on inner or outer perception. Meinong classifies them as a priori judgements. This is justified by his theory of higher-order objects. Higher-order objects exist *per se* when the objects of lower order, upon which they are grounded, exist. Relations are analysed as objects of higher order, but also objects like number, size and shape which are taken by Meinong to be complex objects.

To be sure, Meinong's presumptive evidence is weaker than the evidence for certainty. It does not guarantee the correctness of judgement. Judgments with that kind of evidence are only presumptions or surmises.[7] They are not correct with certainty but only with a high probability. Nevertheless, they are cognitively better than judgements lacking any evidence and they are in a certain manner incontrovertible.

5 Meinong's Analysis of Perception

Meinong literally complies with the German term *Wahrnehmung*, which contains *wahr* ("true"). He acknowledges only true perceptions as perceptions and calls false "perceptions" "aspects". An important idea underlying Meinong's view of perception is that all perceptions are judgements. Like Brentano, Meinong takes it that a judgement presupposes a presentation (*Vorstellung*). The former is about the object presented. What is judged is an objective (*Objektiv*, a Meinongian term which was new in German and which does not have the customary meaning of the English word). All perceptual judgements have existential objectives, according to Meinong. The presented object is judged to exist. Predicative judgements are all not perceptual, according to Meinong, because he holds that they all are a priori. He takes for granted that perceptual judgements are never a priori.

The distinction between presentation and judgement is, of course, traditional. The distinction occurs also in Brentano but Meinong departed from Brentano's after he discovered the category of "objective". In Brentano, judgments are responses to presentations, either rejection or approval, while in his mature view Meinong maintains that judgements are intentionally related to objectives. As to perceptions, Meinong holds that all are intentionally related to existence objectives. Meinong's predicative judgements are all intentionally related to objects of higher order and are therefore a priori.

Meinong argues that outer perceptions are evident surmises.[8] Thus they have some evidence although it is an evidence of the weaker kind. Hence, according to Meinong's ontology of knowledge, outer perception is vindicated.

Meinong first held a Kantian view of our knowledge of the physical world. According to this view, we cannot assert anything about the physical world, except that it is the partial cause of our mental phenomena. That was before Meinong adopted the distinction between act and object, which comes from Brentano,

7 Meinong 1906, p. 70.
8 Meinong 1906, p. 89f.

before he introduced intentional relations, and above all before he discovered objectives. Applying the Kantian distinction between phenomenon (what is in the mind) and noumenon (what is in the external world), the later Meinong claims that we know to a certain extent the noumenon via the phenomenon. Nevertheless, he still accepts the argument from physics and holds that the qualities we perceive are not properties of the material objects (they are mere phenomena). But he is now in the position of defending the view that we can transfer the relations of comparison holding between the phenomenal qualities – and also between other objects of higher order such as shape and number, which are build upon those phenomenal qualities – to the noumenal qualities. He describes the transference of relations of comparison from the phenomenal to the noumenal as a half-perception (*Halbwahrnehmung*).[9] Moreover, the later Meinong argues that we also perceptually know the individual noumenal substances. He claims that our perception of individual noumenal substances is evident. Therefore, he takes it that we perceive those substances directly instead of having to know them by transference from the noumenal substances.[10] As I mentioned, Meinong had originally followed the empiricist tradition in dissolving the substance into a complex of quality tropes.

The rules Meinong set for the transference from the phenomenal to the noumenal are unmistakably adopted from psychophysics, although he does not mention it. As a psychologist, he was, of course, familiar with psychophysics, and wrote a paper about it.[11] There is, for example, the rule of the prerogative of difference over equality—a rule which has to do with thresholds. Moreover, Meinong's rule of equality of the forum of perception (that is the equality of initial conditions on both sides of a comparison) can be derived from psychophysics, and from sensory psychology in general, by taking into account the phenomenon of adaptation.

In connection with his concept of presumptive evidence, Meinong has a philosophically important insight which captures how sciences actually proceeds. On his view, we need no absolute certainty. Relatively better evidence suffices.[12] Empirical sciences progress by choosing the respectively better data. That is at least one way to advance.

As I already mentioned, Grossmann argues that Meinong could have arrived at an unrestrictedly realistical analysis of perception if he had seen the weakness

9 Meinong 1906, p. 94.
10 Meinong 1906, p. 91f.
11 Meinong 1896.
12 Meinong 1906, § 24.

of the argument from physics (see Grossmann 1976) and abandoned it. I do not agree because one can show that Meinong would have faced the same difficulties, and could have resorted to the same rules to solve them, even if he had assumed that the qualities we perceive are the noumenal qualities.

I did that in a book criticising standard views in philosophy of science, especially the analysis of comparative relations. I came to the conclusion that those relations are not immediately given but have to be inferred. They have to be inferred, on the one hand, on the basis of a theory of the various kinds of properties and their structure, and on the other, on the basis of theories about the comparative techniques applied. The former corresponds to Meinong's a priori knowledge of the relations structuring continua.[13]

References

Bergmann, Gustav (1967), *Realism. A Critique of Brentano and Meinong*, Madison (Wisc.): University of Wisconsin Press.

Brentano, Franz (1970), *Versuch über die Erkenntnis*, Hamburg: Meiner.

Grossmann, Reinhardt (1974), *Meinong*, London: Routledge.

Grossmann, Reinhardt (1976), "Perceptual Objects, Elementary Particles, and Emergent Properties", in: H. N. Castañeda (ed.), *Action, Knowledge, and Reality. Critical Studies in Honor of Wilfrid Sellars*, Indianapolis: Bob Merrill, p. 129–146.

Meinong, Alexius (1896), *Über die Bedeutung des Weberschen Gesetzes*, Hamburg/Leipzig: Voss.

Meinong, Alexius (1902), *Über Annahmen*, Leipzig: Barth.

Meinong, Alexius (1906), *Die Erfahrungsgrundlagen unseres Wissens*, Berlin: Springer.

Tegtmeier, Erwin (1981), *Komparative Begriffe. Eine Kritik der Lehre von Carnap und Hempel*, Berlin: Duncker & Humblot.

Tegtmeier, Erwin (2014), "Realism and Intentionality", in: G. Bonino, G. Jesson and J. Cumpa (eds.), *Defending Realism. Ontological and Epistemological Investigations*, Berlin: De Gruyter, p. 247–264.

13 See Tegtmeier 1981.

Mauro Antonelli

Vittorio Benussi's Genetic-Experimental Phenomenology of Perception and its Place in the History of Gestalt Psychology

Abstract: For a long time, starting from Boring, Vittorio Benussi's position and role in the history of phenomenological psychology, and more generally in the history of psychology, has been prejudicially under-appreciated. He was considered a follower of the Graz School, whose scientific contributions, while relevant, were destined to be absorbed into the much more significant results achieved by the Berlin Gestalt school. This essay aims to radically review this stereotyped image, highlighting the originality and actuality of Benussi's theoretical and experimental project. The latter combines a phenomenology of the perceptual field, aiming to highlight, in agreement with the gestaltist theoretical framework, its autochthonous structuring according to specific principles or laws, with a genetic phenomenology, designed to unveil the temporally structured genesis of intentional conscious experiences, and their stratified unfolding. These experiences, in fact, are organised on several levels, based on one another, and not merely juxtaposed.

1 Introduction

About 1914 Vittorio Benussi was unanimously considered to be one of the most rigorous and brilliant experimental psychologists of his time. His scientific activity had been characterised by an impressive experimental research program devoted to optical-geometrical illusions,[1] Gestalt perception,[2] time psychology,[3] apparent motion,[4] psychology of comparison,[5] and psychology of

1 Benussi 1902, 1904, 1905.
2 Benussi 1904, 1906, 1907a, 1909, 1911, 1914a.
3 Benussi 1907b, 1908, 1913.
4 Benussi 1912, 1915, 1917, 1918.
5 Benussi 1907b, 1908, 1910.

Mauro Antonelli, Department of Psychology, University of Milano-Bicocca and Department of Arts and Philosophy, University of Trento
mauro.antonelli@unimib.it

https://doi.org/10.1515/9783110664850-011

testimony[6] published in the most prestigious German journals of experimental psychology.[7]

Especially in the field of Gestalt psychology, Benussi had achieved extremely significant results from the early twentieth century, which were converging, though not entirely congruent from a theoretical point of view, with those achieved during the 1910s by the Gestalt psychologists of the Berlin School. For this reason, in 1915, in a long and dense essay, Kurt Koffka[8] vehemently confronted Benussi's theory of perception and more generally the theses of the Graz School, to which Benussi belonged. Koffka's criticism, though partially founded, appears to be really ungenerous and aimed merely at polarising their respective positions in order to show that Benussi was ultimately anchored to the old atomist and associationist models based on the constancy hypothesis, which only the Berlin school had definitively overcome.[9]

This controversy significantly contributed to the obliteration of Benussi's reputation and work for a long time, so that his position and role in the history of phenomenological psychology, and more generally in the history of psychology, have not been, until today, appropriately recognised. Starting from Boring,[10] the conventional image still largely dominating historiography and handbooks is that of the follower of the Graz School, whose scientific contributions, while relevant, were destined to be absorbed into the much more significant results achieved by the Berlin Gestalt school. This essay aims to radically review this stereotyped image.

2 Benussi's Early Research on Gestalt Perception

In his early research on Gestalt perception, Benussi, under the influence of the "production theory" of his teacher Meinong,[11] came to distinguish two orders of

6 Benussi 1914b.

7 For an overall exposition of Benussi's work, see Musatti 1928; Antonelli 1994, 1996, 2005, 2006, 2018; Sinatra 1998; Cattaruzza 1999.

8 Koffka 1915.

9 Antonelli 2005, 2018.

10 Boring 1929, 1950.

11 According to Meinong (1902, § 2; 1910, § 2), two or more presentations of elementary (real) objects (inferiora, e.g., sounds or points) can produce the presentation of a complex or higher-order (ideal) object (superius) (such as a melody or a spatial figure). Whereas the lower-order objects are the *foundation* of the higher-order object, the presentations of the elementary objects *produce* the presentation of the superius. The "production of presentations" (*Vorstellungs-produktion*) is therefore a concrete mental activity, and not an ideal and static relationship which is

presentations: presentations of "sensory" and "non-sensory origin".[12] The presentation of a sound or of a colour are examples of the first, while the presentation of a complex object such as a difference, a melody or a spatial figure are instances of the second. While the former is exclusively determined by the sensory stimulus conditions, the latter gives rise, by constant stimulus situation, to various perceptual outputs. In other words, while presentations of sensory origin are based upon purely physiological processes of receiving, transmitting and processing of sensory impulses, the presentations of non-sensory origin are closely dependent on particular "attitudes" (*Verhaltensweisen*) or "consciousness settings" (*Einstellungen*) assumed by the subject, which intervene organising and structuring the sensory data. This distinction was reached by analysing cases of "Gestalt ambiguity" (*Gestaltmehrdeutigkeit*), that is, perceptual Gestalten the organisation of which is influenced by the perceptual attitudes of the observer.[13]

For Benussi, thus, the Gestalten are "labile", "restless", "unstable", "ambiguous" (*mehrdeutig*) configurations, and this Gestalt ambiguity is paradigmatic of reversible configurations: here a variable perceptual output corresponds to a state of constant stimulation of the sense organs: one of the two phenomenal objects remains 'invisible', even though the objective conditions for its perception are all given, so that the subject, maybe through appropriate suggestion, suddenly sees it, without anything changing in the physical and physiological conditions of her perception.

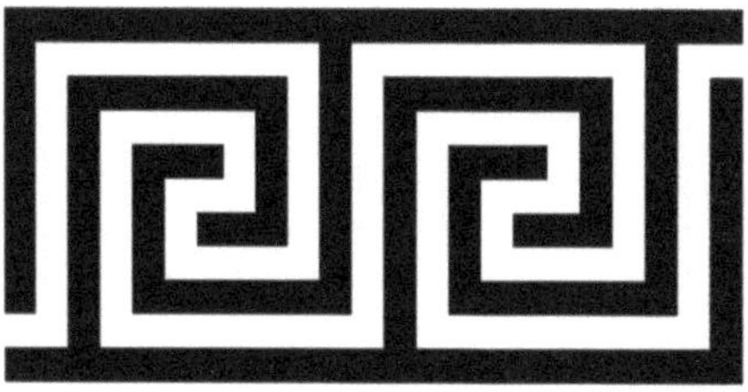

Fig. 1. Meander (from Benussi 1914a, Fig. 1)

the expression of an ontological, formal a priori. It is a construct that highlights the distance from classical associationism; the "production" responds to its own peculiar criteria, which are inter alia independent on the establishment of habits and require a certain 'quantity of work' by the mind. The production theory, only sketched by Meinong, will be elaborated in more detailed way by his pupils Rudolf Ameseder (1904) and Stephan Witasek (1908).

12 Benussi 1904.

13 Benussi 1904, 1906, 1907a, 1914a.

If we observe the complex labyrinth shown in Figure 1, it may appear as "a white figure that develops endlessly on a black background", or as "a figure formed by two series of hooks opposite to each other on a white background".[14] These two aspects are phenomenally heterogeneous and irreducible to one another; their alternating is accomplished without the slightest change in the stimulus conditions.

Fig. 2. The Zöllner illusion

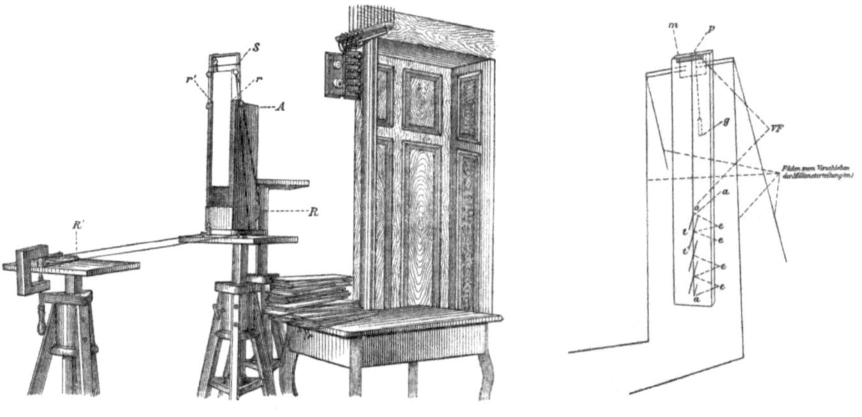

Fig. 3. The experimental apparatus designed and built by Benussi for the study of the Zöllner illusion (from Benussi 1902, Fig. 2 and 3)

14 Benussi 1914a, p. 398 [343]. Benussi systematically studied reversible configurations from 1904 onward, well before the well-known research by Edgar Rubin (Rubin 1915a; 1915b).

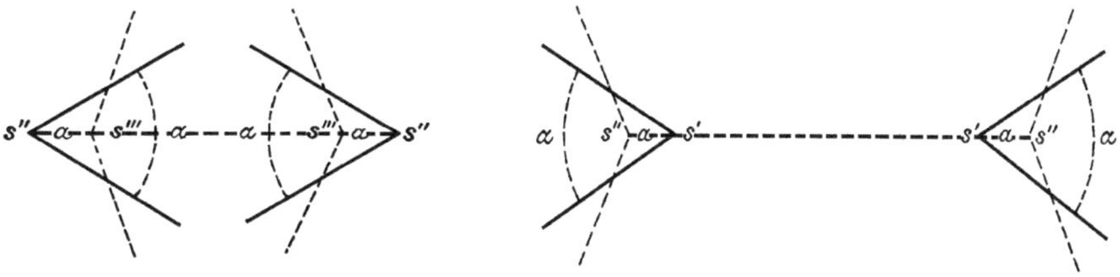

Fig. 4. The Müller-Lyer illusion (from Benussi 1904, Fig. 8)

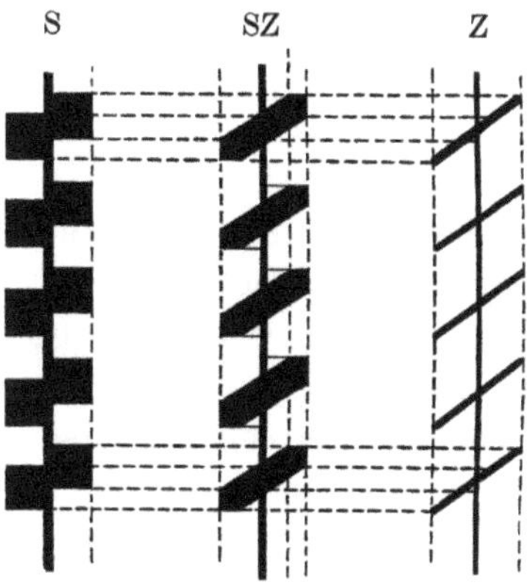

Fig. 5. The chess-board illusion (from Benussi / Liel 1904, Fig. 2)

This ambiguity applies not only to reversible configurations but also to optical-geometrical illusions (Figs. 2–5): the more attention is focused on the single parts or elements of the figure (which are of sensory origin), the less one is subject to the illusory effect of the whole Gestalt (of non-sensory origin). Illusions arise when a setting of the synthetic type prevails; the assumption of an analytical approach, on the other hand, leads to a decrease in the illusory effect, and even to its annulment.[15] Even simple complexes of points or sounds, arranged in a relatively uniform structure or sequence, may give rise to different visual forms or melodies, depending on the way they are perceptually organized in spatial forms or musical phrases.[16] And the same applies to the subjective deformations characterising the perception of duration and temporal flow.[17]

15 Benussi 1904.
16 Benussi 1906a, 1907a, 1909.
17 Benussi 1913.

Benussi's principle of Gestalt ambiguity implies a review and reformulation of Meinong's and Ameseder's *Produktionstheorie*, with a shift of emphasis from the question of the ideal nature of the Gestalten to that of their equivocity: Gestalt ambiguity becomes the core of the theory, and the relationship between inferiora and superiora shifts to the level of precise experimentation. A similar shift of emphasis concerns the subjective activity involved in the constitution of higher-order objects, so that one can speak, more accurately, of certain mental attitudes influencing the organisation of the basic contents in specific ways. The latter can also be favoured by certain traits of the observer's personality (synthetic vs analytic), which, in a sense, act as automatic assumptions for certain attitudes.

In order to avoid misunderstandings, it should also be emphasised that Benussi's elementary data (which Meinong had defined "inferiora") are not to be intended as physical or physiological stimuli, or as the alleged elementary sensations corresponding to these stimuli. The elements of the perceptual foundation (*Grundlage*)[18] are not located beyond direct experience: they are the basic elements present in the observable structures, without which these same structures would cease to subsist. Given these assumptions, the focus here is not so much on the mechanisms operating between stimuli and perception, but rather on the relationship between the elementary sensory objects and the global structures emerging from them, which are both equally present at the time of observation. On the other hand, given the theoretical paradigm inherited from Meinong, and Benussi's substantial loyalty to the methodological canons of experimental psychology of his time, he was to a certain extent obliged to work on very complex levels: the level of stimuli, of elementary presentations and of the corresponding sensory objects, of the non-sensory presentations and of the corresponding Gestalt phenomena. He also varied the objective stimulation, as well as the observers' attitudes, although never forgetting the distinction between the levels themselves. In fact, the physiological level was of little importance to him, and he only referred to it in certain specific situations.

Benussi's analysis thus considered the issue of the subjective involvement in constituting the givenness of experience. When observing organised objects, we always see the elements *in* the structures, the inferiora *in* the superiora. But by taking a synthetic or gestaltic attitude, the elements are changed by virtue of their

18 With the term *Grundlage*, Benussi means the objectual basis or foundation (Meinong's inferiora), on which emerge multiple and different Gestalten, by virtue of otherwise productive acts. The term goes back to Ehrenfels (1890), who used it to illustrate his concept of "Gestalt quality".

being placed within a whole, within the superius. The elements found the structures, but the latter, by virtue of the mental *acts* constituting them, exercise a feed-back on the elements, *modifying* them.

Here Benussi definitely went beyond the theses of his teacher Meinong, according to whom an element taken in isolation and an element grasped within a structure are absolutely identical. Thus, for Meinong the analysis, which was considered to be the decomposition of experience into its elements, did not lead to an alteration of the analysed content, but merely made it available or noticeable to consciousness. The same datum, therefore, could for Meinong present itself in two different forms, it could be noticeable or unnoticeable, but without being affected in its essence.[19] Therefore, like Ehrenfels,[20] Meinong maintained the hypothesis of constancy, and questioned the conception of perception as being made up of composition-units. However, he explored the manner in which these units are aggregated, as well as the relationship between inferiora and superiora. He referred to the principle that at the elementary level of experience there are elements, and that complexes can always be analysed into their simple units.

Benussi's approach, when considered from this specific point of view, did not only overcome the conceptions of Meinong and the other members of the Graz School, but furthermore did not differ greatly from the working model of the Berlin gestaltists. It is thus incorrect to claim that Gestalt theory overthrew Benussi's (and the Graz School's) epistemological approach, by showing that the inferiora do not found the superiora, that elements do not determine structures, but rather that the properties of the whole regulate those of the parts.

3 Benussi vs. Berlin Gestalt Psychology

Beyond certain mutual misunderstandings, Benussi's meeting with the gestaltists was however destined to end controversially. Koffka's main criticism concerned Benussi's distinction between presentations of sensory and non-sensory origin, which he interpreted as an adherence to the atomistic and associationist models, based on the constancy hypothesis, of much of the previous psychology.[21] In reality, the controversy seemed to be motivated by a clear intent: the

---

19 Meinong 1894.
20 Ehrenfels 1890.
21 Koffka 1915, p. 15–45.

"project of using Wertheimer's Gestalt theory experimentally" necessarily required a firm stance against "Benussi's research, some of which concerns the same issues", and which "was developed under the guidance of a well-defined theory, very different from Wertheimer's, namely the production theory developed within the Graz School". Hence the need "to compare the two theories in a comprehensive manner" and to "highlight their different relationship to experimental issues".[22]

Gestaltists' experimental phenomenology excludes in principle any hierarchy within the domain of objects. There are no founding and founded objects, real and unreal, existing and subsisting objects, arranged according to specific foundational relationships, but an autochthon articulation of the perceptual field, where the dependence relationships develop in many directions: elements and structures entertain part-whole relationships, in which, if anything, the larger structures condition the parts, even though not in an absolute manner. In addition, the field model of Gestalt theory does not consider the subject an absolute entity, independent of the objects around him. The phenomenal self is itself a moment or inner articulation of the total field of experience, and does not have thus absolute consistency or clear boundaries, but takes shape from time to time, depending on the overall field situation, and is subjected to the same laws that regulate the formation of perceptual units.

According to the gestaltists, the articulation of the perceptual field takes place by virtue of a spontaneous distribution of the forces present in it. Starting from this assumption, Wertheimer showed how the formation of the phenomenal units obeys autochthonous principles or factors, the so-called "Gestalt laws":[23] proximity, similarity, closure, continuity of direction, common fate, etc. These laws can in turn be traced back to a more general principle: perceptual organisation tends towards the state of Prägnanz, to a "good" form, to the "best", "most sensible" or "more coherent" configuration. Perception obeys the principles of simplicity and economy, tending spontaneously towards a maximum balance between the forces at play, and therefore to a maximum of stability and resistance to change.[24] Underlying this assumption is of course a very specific brain theory, according to which brain processes tend to produce a maximum of homogeneity, compatible with the stimulation conditions at the periphery of the nervous system.[25] The degree of Prägnanz varies within precise

22 Ibid., p. 14.
23 Wertheimer 1923.
24 Ibid.
25 Köhler 1920.

limits, depending on whether the autochthonous structuring forces of the perceptual system can develop more or less completely.

It is this principle that characterises the Gestalt approach to the study of perception: the perception system is a complex dynamic system, a *field*, in which the distribution of the forces proceeds in an autonomous and self-regulating manner. The concept of modification plays a decisive role in this theoretical framework: there is no element in the perceptual field that is entirely isolated from the rest and, as a consequence, all elements are modified by, and in turn modify, the other perceptual components (*Ganzbestimmtheit der Teile*).[26] These modifications, however, instead of being related to the apprehensive attitudes of subjects, to the attitudes of their perceptual modes, are immanent in the system, referring back to its autochthonous organisational laws. Gestalt theory did not deny the influence of 'external' unification principles, such as attention or habit. However, it did firmly deny that these are the only, or even the most important forces that structure the perceptual field. In particular, the interplay of these variables is attributed to the overall action of field factors, i.e., to their global functional unity.[27]

Thus, the intentional model of consciousness inevitably receded into the background. It did however not disappear altogether, since Gestalt theory maintains, at least as a first approximation, a relational or systemic conception of the field of consciousness: the perceptual field is a complex whole, articulated into two subsets, respectively relating to the phenomenal environment and the phenomenal self. These two subsets are in a relationship of functional interdependence.[28]

---

26 Metzger 1954, p. 75–95, 112–118, 258–270, 314ff.

27 Ibid., p. 82: "Gestalt theory is [...] not at all 'opposed to attention'. Careful readers cannot have had the impression that it regards subjects or their organs of perception as a 'tabula rasa' or a wax plate, into which stimuli simply 'imprint' something. [...] Rather, in this theory, the effects of attention, of the direction of observations, of modes of apprehension, have a very definite place, as special cases of the effect of the whole on the parts (*Ganzbestimmtheit der Teile*). Again, it merely opposes the unjustified claims of theories of perception based on attention, observation, and comprehension, which are not to be preferred to other, equally possible theories, without first thoroughly examining the domains in which they are valid."

28 Koffka 1935, p. 147: "In the first place, under certain circumstances, we find the field organisation dependent upon attitudes, i.e., forces which have their origin not in the surrounding field at all, but in the Ego of the observer. This is a new indication that our task of investigating the surrounding field alone is somewhat artificial, and that we shall understand its organisation completely only when we study the total field which includes the Ego within its environment." Cf. ibid., p. 319ff., 519ff., 594ff.; Köhler 1947, p. 123ff.; Metzger 1954, p. 81ff. and Ch. VIII.

A consideration of the concrete experimental research of Gestalt psychologists, however, seems to lead to different conclusions: the impact of subjective factors seems to be a mere footnote, while the role of the so-called structural or autochthonous factors is emphasised. This attitude depends, at least partly, on the initial strategy of the Gestalt theorists, who opposed the traditional theories in which subjective factors were considered to be the most important factors of perceptual organisation, if not the only ones. In reality, rather than denying the impact of the subjective, which is after all undeniable, the gestaltists limited themselves to reducing its scope of influence. When one considers the question of the significance of objects, of their 'imposing' or 'giving themselves' to the subject, in a word their *Auffälligkeit*,[29] the alleged objectivism of Gestalt theory can be somewhat overthrown, by saying that if the subject adopts the right attitude, then things 'impose themselves', or 'give themselves' to the subject.

The two theoretical models, however, are not diametrically opposed; rather, they are complementary models that can integrate and complement each other. The work method of the gestaltists did not greatly differ from Benussi's, when one leaves aside the 'philosophical' character which Gestalt theory adopted at times, and considers its more authentic nature as a theory generating empirical research, as an experimental phenomenology of phenomenal reality.

Of course, in Gestalt theory a tendency prevailed to explain the phenomenal with the phenomenal, to establish a set of functional relationships between the various aspects or moments found in the phenomenon under observation, which 'reabsorbed' the interplay of subjective or non-sensory factors. But in these explanatory dynamics, it is always possible to shed light on the role, albeit a background one, of the subject, on whom these aspects or moments 'impose', 'give' themselves with absolute necessity. At these moments, it is also possible to show the privileged role played by certain aspects or phenomenal moments, which have, as it were, the task of generating or 'founding' other ones – without undermining the general assumption that the properties of the whole determine those of the parts.[30] In this sense, one can also find the presence of founding and

29 Metzger 1954, p. 188ff.

30 This is particularly evident in the work of the Italian Gestalt psychologist Gaetano Kanizsa (1913–1993), a pupil of Cesare Musatti (Kanizsa 1976, 1979,1980, 1991). Kanizsa proceeded to submit visual structures to his experimental subjects, and then to modify them systematically in order to obtain unexpected and often paradoxical perceptual effects. All this was based on a percept-percept coupling process, according to which "a phenomenal property acts on another phenomenal property directly and visibly", whatever the state of the corresponding stimuli (Bozzi 2003, p. 192). In his procedure stimuli and perceptions were not properly at

founded objects, of inferiora and superiora in the explanatory models of Gestalt theory. The postulate of the dependence of the parts on the whole is nothing more than a convenient shorthand abbreviation, in itself vague and abstract; each situation must be studied individually, in order to concretely establish the role of the factors that are at stake at any given time.[31]

If one considers the concrete research practice of the gestaltists, the general categorical statements give way to less unyielding statements, which often represent concessions to the views of their opponents. In fact, much of their experimental material and their field of research is almost identical to Benussi's. The research methodologies are both founded on a phenomenological attitude, understood as the exclusive attention to the perceptual facts as they are given in direct experience. Benussi's review of his previous theoretical model during the period of Padua approached the positions on a theoretical level as well.[32] The only distinction seems to be the interest of the Berliners for what lies 'behind' the perceptual facts, an interest, however, that in the further developments of the Gestalt movement would increasingly recede into the background.

4 Benussi and Husserl

In another point Benussi went beyond the Graz School and the Brentanian tradition, as well as much of contemporary psychology, also in this case as a consequence of his experimental research on "Gestalt ambiguity". This concerned the judgmental nature of perceptual phenomena, sustained by Meinong and Brentano.

Perception has for Benussi its own specific objectuality, which consists of a final and not further analysable objectual element: the *presence* itself. In "perceptual" or "real presence" the object 'is there', 'in front of us', in an immediate

play, as in the classic experiments of the Berlin School, but inferiora and superiora both well present at observation. This method brought the gestaltist Kanizsa close to the Graz School of Meinong and Benussi.

31 Bozzi 1980.

32 In the face of Koffka's criticism Benussi was induced during the Padua period to reduce the scope of his early dualism of sensory vs non-sensory presentations, preferring to speak of outer (or peripheral) vs inner (central) conditions of perception and recognising their blended contribution in every perceptual event. Among the inner conditions of perception, Benussi placed, beside the "figural" or "Gestalt functions", "assimilative functions", which are rooted in past experience, within a theoretical project designed to reconcile the structural-formal point of view with the empirical one.

and intuitive manner. This feature of presence may concern both external objects and objects, events or states of our inner life; one can then speak, respectively, of an "external perceptual presence" or an "internal" one.[33] An object, on the other hand, may be present to the subject without being an object of perception; the object is in this case "perceptually absent" and, nevertheless, it is "representatively present" (a "presence of unreality"). "To experience the absence of an object" – Benussi writes – "is to represent it".[34] In representative states, we are confronted with the objectual feature of "a mental presence only" or a "presence-absence".[35] Compared to this presence, whether it is perceptual or representational, the constative or judgmental attitude, and thus the actual existence of the object implied in the perceptual context, becomes irrelevant. This cognitive integration – any cognitive integration – is inessential to the object of experience with its perceptual qualities, namely with its constitutive properties, available to phenomenological analysis.

In 1926, in an unpublished brief note, Benussi knowingly emphasised his irremediable departure from Meinong's thesis in this regard, and placed his concept of perception in direct relation to that developed by Edmund Husserl in his *Logical Investigations* (1900–1901):

> [My concept of perception] is a descriptive concept that aside from everything that relates to the conditions of perception, is very similar to that of Edmund Husserl (*Logische Untersuchungen*), and it is in sharp contrast with that of the School of Brentano, Meinong, Höfler, and Witasek.[36]

For Husserl, as for Benussi, in perception the object is in fact present "in itself", "immediately", "originally", "in-flesh-and-blood", and just the purpose of freeing psychological-descriptive accounts of perception from their subordination to the problem of knowledge is an element that brings together Husserl and Benussi's phenomenological analyses of perception.

Another common element is the analysis of the intentional acts by virtue of which the chaotic mass of sensory atoms crystallises in a meaningful structure. For Benussi, as for Husserl, the perceptual act is in fact an "apprehensive act" (*Auffassungsakt*), through which consciousness gives an objectual and unitary value to single sensed moments. As such, it cannot be explained on the basis of

---

33 Benussi 1925a, p. 10–12 [220 f.]. Cf. Benussi Archive, Lezioni 1926/27, Lesson 9, box 12, file 7; Benussi 1922–1923, p. 58.

34 Benussi 1925b, p. 36 [192].

35 Benussi 1925a, p. 10–12 [220 f.].

36 Benussi Archive, Notes, box 16, file 22.

the purely physiological mechanisms of reception, transmission and processing of sensory inputs. Rather, it refers to processes which are, in a strict sense, non-sensory (apprehensive or apperceptive functions according to Husserl,[37] special comprehending settings of the subject, according to Benussi), and which intervene by organising and structuring sensory data.

It is well-known that, according to Husserl, sensory data and sensations were not intentional; for a reference to an object to develop on the basis of a complex of sensations, these must be "apprehended" objectually. This "apprehension" (*Auffassung*) or "apperception" (*Apperzeption*) leads to a "surplus" (*Überschuß*) that is essential to the constitution of the object.[38] The term 'surplus' alludes to the fact that the intended object is always 'more' than the actual content of consciousness, the raw material of sensations. Thus, in the case of the perception of an object, only a part, an aspect or adumbration (*Abschattung*) of it is actually given intuitively. The aspects that are not given to intuition, which belong to the "surplus", are understood in a purely imaginative or symbolic manner. The object in its entirety is only available in an intuitive progress, in which the imaginative and signitive intentions are gradually fulfilled. This development obeys an essential law, which depends on the essence of the corresponding object. Conversely, the same complex of sensory data can lead to the perception of different objects, on the basis of different apperceptive interpretations.[39]

Husserl's thesis fits perfectly with Benussi's principle of Gestalt ambiguity: variable perceptual output corresponds to a state of constant stimulation. Benussi attributed the latter to non-sensory factors, Husserl to different apperceptive interpretations of the same sensory complex. There is a parallel between subjective and objective, noetic and noematic: it is essential for a thing to have a hidden side; it is in the nature of perception that this side is somehow also perceived. The role of non-sensory acts, the issue of the constitution and the parallelism between noetic and noematic levels are the core themes that place Benussi's analysis in close proximity to Husserl's.[40]

37 Husserl 1900–1901.

38 Ibid., II, p. 385; Eng., II, p. 567.

39 In this regard, Husserl introduced the example of unstable perceptions, that is, perceptual situations in which, faced with the same set of sensory data, a conflict develops between two different apperceptive interpretations of the same complex (mannequin – real person) (ibid., p. 442 ff.). Husserl used this example on several occasions; cf., e.g., Husserl 1948, p. 99 f.; Eng., p. 92 f.

40 Benussi's work – especially his research on time consciousness – did not escape Husserl, who kept a copy of Benussi's *Psychologie der Zeitauffassung* (Benussi 1913; cf. Hoofdwoorden-cataloog – 09/04/87) in his library. Benussi is also expressly mentioned in some of Husserl's manuscripts (D 13 III / 208a; K II 4 / 59–62; BQ 27 / Vorbl.). Cf. Schuhmann 1975.

But the similarities between Husserl and Benussi's projects extend beyond a mere static analysis of consciousness, referring to the common project of a genetic phenomenology. It is designed to unveil the temporally structured conscious genesis of intentional experiences, and their stratified unfolding, i.e., those experiences that are organised on several levels, based on each other, and not merely juxtaposed.

Of course, beyond the often very surprising affinities between Husserl and Benussi's projects, a fundamental difference remains: while Husserl's genetic phenomenology programmatically intended to be an eidetic investigation, designed to bring out the necessary stages through which intentional experience is formed, Benussi's project was an empirical and experimental investigation of the stages or phases, that can be determined empirically, through which our consciousness of objects is constituted in time. It made use of the instruments available to the experimental psychologist at the time: tachistoscopic presentations of visual and acoustic stimuli of different temporal amplitudes, increases or decreases in the intensity and brightness of the elements of Gestalt complexes, increases, reductions and rotations of visual Gestalten – techniques which were of great importance to cognitive psychology during the last decades. The use of these techniques aimed to analyse the *Teil-* or *Vorgestalten*, characterised by a relative mutability in the relationship between their parts, and their tendency to gradually take on a more stable shape.[41]

5 Genetic Psychology

Thus, for Husserl the phenomenon exhibits two characteristic levels: on the one hand, it shows itself immediately for what it is, as something that is directly grasped in the here and now, and not construed according to subsequent structuring processes; on the other hand, it refers, for its constitution and explanation,

41 Benussi's research on *Gestaltzeit* (Gestalt Time) (Benussi 1914a; cf. Benussi Archive, Die Gestaltzeit, 1912, box 4, files 2, 3, 4, 6, 7, 9) should be placed in this framework. The *Gestaltzeit* being the time required by the apprehensive act in order for the sensory data to organise themselves gradually into (relatively) stable percepts. Benussi used figural complexes attributable to well-known illusory figures and tachistoscopically projected some of their figural elements (e.g., the slanting segments of the Müller-Lyer figure) with different exposure times (brief, indifferent, long), while assessing the exposure time necessary for the occurrence of the maximum illusory effect, or for the full implementation of a grouping or Gestalt attitude. He discovered that this time was variable, depending on the constitutionally analytical or synthetic attitude of the observer.

to a series of successive apprehensions, since it is only in a temporally structured development that the consciousness of what is currently given is possible: what I hear and see is an organised structure, and not a simple aggregate of sensations to which, thanks to the astonishing operations of consciousness, a form is added from the outside. The formation of any structure requires continuous connections in the stream of consciousness, through the retention of the just passed and the protention of the future outcome, associated with the corresponding (positive or negative) fulfilment of intentions following one another in the immanent temporality of consciousness. Behind the directly intuited phenomenon (this figure, this melodic piece, etc.) lies a synthetic work, temporal in nature, made of retensions, protentions, but also of remembrances.

This synthesis is first and foremost passive: it is the expression of *passive activity*, of an obscure and unconscious functioning which necessarily precedes the active synthesis through which consciousness is aware of its constitutive acts. It is clear that the term 'consciousness', which refers to this basic synthesis, becomes inappropriate, as do the terms 'I', 'Ego' or 'subject'. This constituting and original layer of experience is the field of genetic phenomenology, which the terms 'conscience', 'Ego', and 'I' only qualify metaphorically; it is the field of what Husserl defined a "pre-Ego" (*Vor-Ich*), i.e., as an anonymously self-acting Ego.[42]

At the level of genetic analysis, Husserl thus went beyond the previous distinction (without repudiating it) between the activity of consciousness, which synthesises and animates, and the passivity of matter (*hyle*), i.e., the complex of sensations available to be animated. Originally, genetically, in fact, the *hyle* itself tends to be articulated according to its passive laws; the active constitution has its counterpart in passive self-constitution, in that passive synthesis that obeys the universal formal law of temporality and the associative laws (secondary associations, such as those that explain memory, and original associations which explain the self-structuring of the perceptual field). Behind the transcendental subject, no longer intended as a pure formal structure, but rather as a concrete monad, one discovers the presence-absence of the *Vor-Ich*, which is the genetic origin of the subject and the domain of the passive genesis.

Husserl also defined static phenomenology as "descriptive" phenomenology, since it is the phenomenology of the essential possible forms of consciousness, whereas he defined genetic phenomenology as "explanatory", since it aims

42 Cf. above all Husserl 1966; 1973. Of the large amount of critical literature available, I limit myself to reporting only Kern 1964; Held 1966; Brand 1969; Holenstein 1972; Yamaguchi 1982; Miller 1984; Rang 1990; Römpp 1992; Kühn 1998; Taguchi 2006; Sandmeyer 2009; Biceaga 2010.

to identify the genetic laws of the original constitution. Genetic phenomenology thus undertakes a *Rückfrage*, a "questioning-back" aiming to establish all possible ways of the genesis, thus reaching its fundamental principles. From there, through a sort of reversal, Husserl traced the same path in the opposite direction; he started from the genesis of passivity, and returned to the involvement of the Ego and the relationship between activity and passivity.[43]

Also in this respect the affinity between Benussi's and Husserl's model is truly surprising. For Benussi the activity of "apprehension" (*Auffassung*) of perceptual objects, which to descriptive analysis appear to be crystallised in segregated units, as equipped with their own specific consistency and stability, presents in fact an internal and complex stratification. The necessary condition for its emergence is the establishing of a "real relation" (*Realrelation*) among sensory data, according to which "sensory contents influence each other according to their specific properties".[44] That is, one finds here a connection that does not call into question mental activities capable of instituting it, with which, the subject, therefore, relates in a purely passive way.[45] At this original level of perceptual experience, the segregation of the field takes place automatically, by virtue of the real relations of similarity and difference between the perceptual contents: any sensory datum influences, and is subjected to the influence of all the other data, leading to the grouping of similar elements, and to the isolation of dissimilar ones.[46]

From the perceptual contrast, emerging from the relationships of homogeneity and heterogeneity of the parts, and from the correlative dynamics of the

---

43 He then goes on to the formations of pure actuality, to the constitution of the individuality of the monad, to the inter-monadicity, and the implication of a monad in the genesis of another, to finally conclude with the intersubjective constitution of a single world, which is intersubjectively valid for all subjects. In the 1930s Husserl would call this double, outward and backward movement *Abbau-Analyse* (deconstruction analysis) and *Aufbau-Analyse* (construction analysis).

44 Benussi 1904, p. 394 ff. [90].

45 On the distinction between subjective or ideal relations (produced by the mind), and objective or real relations ("encountered relations") see Meinong 1889. Here Meinong, against empiricism, which tended to trace the foundation of the 'rules' for unifying ideas back to past experience, to a process of accumulation and repetition of impressions that solidify as habits, began to develop the idea that these 'rules' have their basis in the very structure of the perceptual fact. In other words, he began to realise that in addition to subjective or ideal relations, which are the result of a mental activity of comparison or connection of sensory data, some are still immediately – passively – given to the subject, who limits himself to ascertaining them without the intervention of mental activity. For these "found" or "encountered" (*vorfindlich*) relations Meinong used the term "real relations".

46 Benussi Archive, Auffälligkeit, 1905, box 1, file 11.

grouping and the isolation of the elements, the experience of phenomenal *salience* (*Auffälligkeit*) originates, thanks to which the subject, who was initially present only in latency, is involved. Salience, in fact, is not a property inherent to a percept as is colour or brightness, nor is it therefore a property that an object takes on or loses in the same way in which it acquires or loses these types of properties. *Auffälligkeit* is an essentially *relational property*, which cannot be described without the context that binds the phenomenal objects to each other *in their being given to a subject.*[47]

The emergence of a relationship of *Auffälligkeit* within the perceptual field marks the transition from the pre-attentional to the attentional level. Perceiving the 'impact' of the *Auffälligkeit* relationship leads to the mental act of attentively focusing on what attracted it.[48] The involvement of the Ego marks a radical change of perspective: a shift occurs from a situation in which the structuring moments of perceptual experience are all on the side of the phenomenological contents, to a subjective orientation of the entire perceptual field. From a passive observer who looks on the autonomous self-structuring of perceptual contents, the subject is gradually transformed into a living, attentional centre. The tendencies animating and driving perceptual development have their origin in this centre, in particular, the tendencies to the analysis and synthesis of the elements on which the segregation of the perceptual field depends. The Ego, thus, undertakes an active clarification of the connections that emerged from the automatic segregation of the field, assuming, as it were, the 'responsibility' for what was formed independently of its activity.

The *Auffassungsakt*, then, does not, in a strict sense, 'produce' the structure, but limits itself to clarifying it on a higher level: a sensory complex can be "apprehended" according to different directions of meaning, because on the basis of these apprehensive directions, albeit on a latent and passive level, possible structures are already available, which can 'then' be grasped in this or that particular sense.[49]

Benussi's experimental designs should be understood in this light. They maintained the complex of stimuli unchanged, and systematically prescribed variations of the apprehensive modes of the given stimuli. In particular, subjects were requested to vary their tendencies to analyse and synthesise the perceptual elements, leading to a feedback on the related object, with the parallel redistribution of phenomenal salience, and the reorganisation of the perceptual field.

---

47 Ibid.

48 Benussi 1909, p. 73 ff. [229 ff.].

49 Cf. Linke 1918, p. 252–257.

In the afore-mentioned handwritten note of 1926, Benussi emphasised: "The prescribed perceptual behaviours (*vorgeschriebene Verhaltungsweisen*) were 'tasks', at a much earlier date (1904) than the concept of *Aufgabe* of the Würzburg School".[50] Benussi's approach, in fact, was undoubtedly close to that of the Würzburg School. It was indeed this school that induced Benussi to approach the Husserlian theoretical model while studying perceptual processes.[51]

Benussi's investigations, centred around the issue of *Gestaltmehrdeutigkeit*, thus analysed the relational context which connects the objects of experience to one another, and to the subjective source of their arising; to that dynamic, active source, characterised by different degrees of clarity, that transforms the experiential givenness into an object *for* the subject. The increasingly active intervention of the subject involves a gradual transition from the indistinct flowing of the sensory field to the fixity and identity of the object; this process is in fact essentially temporal in nature. Perception qualifies as a temporal development of phases that are unified in the consciousness of the object, each of which is moreover already consciousness of the same object, albeit in a partial and indistinct manner.

6 Time

The principle according to which apprehension, as a subjective act stemming from the Ego, that interprets sensory data, is rooted in a pre-conscious dimension, dominated by laws that are immanent to the field of sensory contents, led, as we have seen, both Husserl and Benussi[52] to a phenomenological analysis of time, as well as to a critical recovery of the concept of assimilation, thus claiming the role played by past experience in the constitution of the perceptual act.

Time was for Benussi a specific dimension of our perception, a dimension that, by virtue of the relatively stable nature of the objects of experience, and of the physical objects that are to some extent their prerequisites, remains mostly hidden, but that should not be disregarded. The very same physical and physiological facts which are at the origin of our experience of objects are not really things, but processes that endure over time. And also from a phenomenological point of view we are confronted with percepts that, strictly speaking, are not things or objects, but more or less stable phenomenal events. Between the two

---

50 Benussi Archive, Notes, box 16, file 22.

51 Cf. Messer 1908.

52 Husserl 1928, 1966; Benussi 1913.

orders of events and their temporal dimensions there is no direct correspondence, of course. On the one hand, in fact, there is a longitudinal series of causally determined physical-physiological events that move 'forward', passing through a series of objectively determined temporal instants; on the other hand, there are subjects who, within their own time frame, transform the punctual instant – the "now" (*Jetzt*) – into the *present*, i.e., structure the inner time according to that principle of organisation that is the experienced, lived present. The units of subjective time qualify, in fact, as intensive and non-independent parts, and as such, they lack a direct correspondence with the extensive or objective parts of mechanical time.[53]

The idea of a temporal articulation of the perceptual act occurs in all Benussi's works and finds its mature expression in his *Psychologie der Zeitauffassung* (Benussi 1913). At the centre of this work lies the phenomenological investigation of inner time, on the structure of duration and its internal relationships, as well as the dynamic constraints which it imposes on the formation of perceptual objects. Benussi starts with an analysis of the "absolute impressions" of time,[54] i.e., those durations that are grasped by the subject in an immediate and intuitive way, regardless of any comparative processes.[55] Moving from the immediate characterisation that experimental subjects provided of simple temporal stretches limited by two sounds, and by observing how the subjects distributed the absolute impressions of time by gradually passing from shorter to longer times, Benussi distinguished five types of "absolute time impressions": *very brief* (from 90 to 234–252 ms), *brief* (from 234–252 to 585–630 ms), *undetermined* (from 585–630 to 1080–1170 ms), *long* (from 1080–1170 to 2070 ms), and *very long* (more than 2070 ms).[56] These absolute "intuitively experienceable" times circumscribed the field of the "mental presence-time" (*psychische Präsenzzeit*),[57] i.e., the

53 Benussi 1913, p. 1–7 [7–11].

54 Ibid., ch. I, p. 9–58 [13–47]: *Die einfachsten Beziehungen zwischen subjektiver und objektiver Zeit* (The Simplest Relationships between Subjective and Objective Time).

55 Ibid., p. 10 f. [14]: "The contrast between intuitive and the non-intuitive can [...] become sufficiently precise by saying: if a mere process of presentation is sufficient for the characterisation of the intellectual experiences which are necessary to make present (*vergegenwärtigen*) an object, then we will be able to call the grasping of this object intuitive; if, on the other hand, these events are insufficient, then we shall speak of non-intuitive apprehension. Thus, one can make present the duration of one second through a mere presentation, while the duration of a year requires a peculiar inner attitude [...], or the thought: 'the time between two limits, which can be determined by means of external limits'."

56 Benussi 1913, p. 15 ff. [17 ff.].

57 Stern 1897.

original temporal field that is characterised by its "immediate nature of perceptual presence, and in which no part of the temporal stretch grasped from time to time appears to be remembered or belonging to the past"; "It can, as a whole, therefore be defined as the 'present'".[58]

Therefore, for Benussi the present firstly had a character of clear perceptual presence. The present is "evident", in the sense that it presents itself in an immediate and intuitive way, i.e., without the intervention of any reconstructive mental function. But this presentation is clear or evident also because it is grasped without requiring any other impressions for its constitution: in the same manner as a time impression can be said to be "absolute" when it is comprehended in its qualitative dimension and regardless of any comparison, the time experienced as "present" is such regardless of being compared with other temporal features, belonging to the past or future. The present, therefore, qualifies as an immediate experiential field, as a temporal field internally articulated into various 'parts', which are all grasped as a whole in direct perception.

The present or "mental presence-time" is, so to speak, the stage on which every real experience takes place. More than an experience which perhaps is only exceptionally experienced, it is the condition of all experience, the condition of the constitution of the unity of objects and of their identity. The identity of the object, its apprehension as an identical pole that lasts in time, refers in fact to the *duration of the apprehensive act*, to its presence-time, by virtue of which past instants, still 'controlled' by the retentional processes, and future instants, anticipated by the protentional ones, give rise to an identification synthesis. This synthesis allows the object to be grasped as the bearer of a *Gestalt identity*, unitary not only with respect to the multiple simultaneously given data, but also to the manifold data distributed in the temporal flow of actual experience.

The objects, in fact, are always pre-given to our observation: gradually entering into our field of observation, they appear as inescapably pre-existing our acts of perception; symmetrically, when they leave the field of observation, they are not abolished when they cease to be seen, but maintain a readiness to be perceived and to be actualised again. The percepts enter and leave the field of observation already having the characteristics of their constitution and continuing to maintain them. The perceptual field is a field of progressive appearance and disappearance of objects in themselves already accomplished, similar to the

58 Benussi 1913, p. 11 [14].

gradual unfolding of a landscape seen from a train window, which is a landscape before we have seen it, and that continues to be a landscape when it disappears.[59]

Time, which in terms of receptivity is a still indistinct stream, carrying on in a latent and passive way, becomes, at the apprehensive level, the residual moment, the kind of background on which the lasting object stands out. In fact, it is the apprehensive level that constitutes the sort of phenomenal split that allows the duration of things to be distinguished from the passive flow of time. Between these two moments an inversely proportional relationship exists: the more phenomenal time is filled with objectual presence, the more this presence, with its *Auffälligkeit*, catches our attention, making time duration as such recede into the background; the more time is empty of objectual presence, the more it assumes a quasi-objectual character.[60]

During the descriptive analysis, the perceptual world appears to consist of stable and motionless objects and properties. In reality, it is made of fleeting events which are constantly transformed into something else, through progressive substitutions and disappearances. In this chaos of punctual events, however, wider dynamic structures are formed than individual instantaneous facts, which become parts or moments of this dynamic Gestalt. The mental present is the unification and identification of a temporally distributed multiplicity, a synthetic form which unifies a plurality of successive elements.

Perceptual experience is made up of successive and yet co-occurrent scenes. This succession within the co-occurrence does not mirror the order of physical time, the objective sequence of stimuli, as is clearly shown by Benussi's research on "temporal dislocations" (*Zeitverschiebungen*) or reversals. This discovery shows that within the presence-time there is an area characterised by the possibility of a reordering of incoming events, that is, in which the functional connections between events can move forward or backward, since in determining the perceived succession the temporal proximity may conflict with other unifying factors, such as qualitative similarity (sound vs noise) or phenomenal salience (intensity of the incoming sounds or colours compared to lesser intense others). This is realised in the core of present-time (*Präsenzzeit*), which Benussi defined *Gegenwartszeit* (whose extension covers approximately 100 ms). The *Gegenwartszeit* thus qualifies as the focal nucleus of the *Präsenzzeit*, which, in turn, unfolds in a temporal halo up to the limits of *Anschaulichkeit*. The *Präsenzzeit* is

---

59 Cf. Bozzi, 1985.
60 Benussi 1913, p. 186 ff. [139 ff.].

thus the original temporal field in which the present perception of time is implemented, which is therefore internally articulated in its duration rather than being instantaneous.[61]

The perceptual process hence develops within a dynamic field, in which the content that is the focus of attention lingers for a certain time, and then turns into the halo that surrounds a new content entering into the mental present, and then gradually disappears. This content may subsequently remain present to consciousness or resurface, but only as a represented (perceptually absent) object, and no longer as a perceptually present object. Of course, the boundaries separating the perceptual present from the "immediate" (or "primary") memory are also unclear and nuanced, as each perceptual fact in the actual present has its shadow, its mnemonic tail, which frays, zig-zagging backwards. This applies not only to elapsed dimensions, but also to those that are incoming. In fact, the properties implicit in a perceptual development, in a defined manner, foreshadow what is still hidden in the present, since these properties can be realised by different sensory aggregates.

7 Assimilation

Therefore, despite the fact that, from a descriptive point of view, there is a clear difference between perceptual and merely mental presences, from a genetic perspective, the two forms of presence qualify as poles of a continuum. This explains how, due to the intervention of assimilative processes, with their roots in past experience, representational elements can be integrated into actually given perceptual elements (*perceptive assimilation*) through a fusion schema.[62] Through the coming alive of assimilative functions, the current perceptual foundation is enriched by elements or features for which the current conditions of stimulation are absent (or insufficient); an example is the apparent three-dimensional nature assumed by certain two-dimensional drawings, such as the Necker cube (Fig. 6).

61 Benussi 1907b, 1908, 1913, p. 272–339 [199–246].

62 In classical psychology, e.g. in Wundt, assimilation means integrating memory-stored information into actual perception; this assimilation can lead to a modification of the actual sensory data to align them with experience-based expectations.

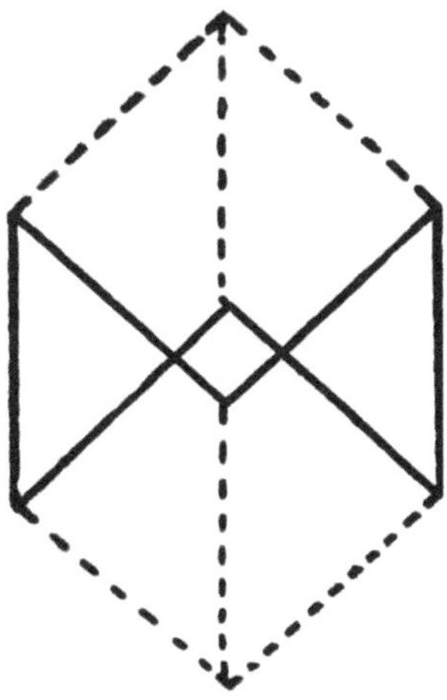

Fig. 6. Necker cube (from Benussi 1912, Fig. 2)

Here not only do functions of completion and integration come into play, but also those of anticipation, which concern the deployment of perceptual scenes and their internal relations, as well as experience in a broader sense, that is the experience of life, the personal history of the observer. Similarly, also the processes of categorisation, recognition, designation and assignment of meaning can be attributed to assimilative factors (*categorical assimilation*). Of course, these processes presuppose already formed percepts, i.e., percepts inserted into precise spatial, chromatic, dimensional and topological relationships. The effect of these factors on perceptual experience is undeniable, but they only exert an *influence*: talking of influence makes sense only where there is already something that can be influenced.

Formal and assimilative factors collaborate in determining the perceptual output, integrating one another in the constitutive act of perception. An unavoidable prerequisite for the arising of assimilative factors is, in fact, the presence of the factors of formal organisation, that secure, so to speak, the basis for the emergence of the assimilative ones, and of the whole perceptual act. The formal factors can, in fact, influence assimilative processes by favouring or hindering them. Benussi's research – begun in 1911 – indicates the emerging of certain forms of apparent three-dimensionality from certain perspective drawings, and the time necessary for these impressions to occur, depending on the type of inclination the prospectively drawn object has.

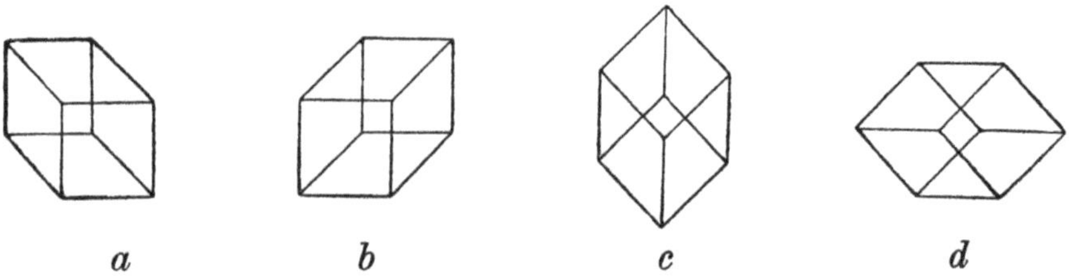

Fig. 7 (from Benussi 1911, Fig. 1)

By orienting a Necker cube according to different perspectives (Fig. 7), one can see how each drawing creates the impression of apparent corporeality employing different times, which Benussi measured by using tachistoscopic methods.[63] The different perspective drawings, and hence the various formal shapes, favour the arising of assimilative processes to different extents, which tend to integrate the formal (two-dimensional) aspect in a three-dimensional sense.

Within the perceptual act a dynamic connection is thus established, by which the assimilative factors are organised and developed starting from the system of structural relationships outlined by the figural ones. The latter are themselves transformed by being inserted into a consolidated system of experience.

In this way, Benussi's experimental investigations showed the presence, in the act of perception and in its presence-time, of at least three constituent layers: 1) an original sensibility, which proceeds according to pre-attentional dynamics in which the Ego is only latently present; 2) the "perceptual apprehension" (*Wahrnehmungsauffassung*) which, as conscious awareness of an objectual presence, consists of making explicit or actively restructuring an already implicitly given structure; and, 3) the identification and categorisation of the percepts, which consists in the enrichment of an already gestaltic structured object.

8 Conclusion

The originality of Benussi's theoretical and experimental project is thus clear: it combined a phenomenology of the perceptual field, aiming to highlight, in terms perfectly congruent with the gestaltist theoretical framework, its autochthonous structuring according to specific principles or laws, with a phenomenology of the subject. Its latent genesis is attributed to the folds of the perceptual mechanism

63 Benussi 1911.

itself – a phenomenology which the Berlin Gestalt school, although guessing its importance, had not been able to promote or develop. In reality, the main difference between Benussi's theoretical-experimental project and that of the Berlin gestaltists is that, whereas Berlin's experimental phenomenology was static, Benussi intended to develop a genetic phenomenology, designed to reconstruct the diachronic laws of the evolution of perceptual phenomena, their meaningful structuring unfolding over time. It was a science of perceptual objects that takes into account not only their current constitution, but also their genesis and their development. It was an experimental-phenomenological model that the gestaltist carefully refrained from pursuing, because by implying a constructivist approach to the study of perception, it did not sufficiently cut its ties with the old, dangerous distinction between sensation and perception, and with the constancy hypothesis, as the 'original sin' of classical psychology in general (cf. Kanizsa, 1952). In Benussi these two moments were instead integrated into a unitary conception, in which the primacy was neither of the subject nor of the percept, but of their correlation. The horizon of his investigations, thus, became three-dimensional: it became an analysis of the implicit operations present in every act of perception, an analysis of the operational network through which the object becomes an *object for* the subject.

Hereby, Benussi demystified the common notion of phenomena as neutral and absolute entities, as independent of the subjective source of their givenness. He rather brought to light the relational structure of experience, articulated on a dual level: not only the 'horizontal' dimension, which regards the intrinsic relationships within the phenomenal domain, but also the 'vertical' dimension, which connects this relational structure to the subjective source of its arising, that is, to the dynamic, operational – though often latent and unconscious – source, which transforms the givenness of experience into an object for the subject.

What then was the real position of Benussi within phenomenological psychology? Benussi was not an out of date precursor, as historiography, starting with Boring, has long advocated. Rather, he embodied the differences, similarities and identities subsisting between Graz *Gegenstandstheorie* and Berlin *Gestalttheorie*, which, in a sort of *Aufhebung*, he surpassed and sublimated in his scientific and personal paths. Benussi was neither a repentant Meinongian philosopher and psychologist, nor an unorthodox gestaltist, but a tireless and curious researcher who, as an *Einzelgänger*, was able to capitalise on that opposition.

References

Ameseder, Rudolf (1904), "Über Vorstellungsproduktion", in: A. Meinong (ed.), *Untersuchungen zur Gegenstandstheorie und Psychologie*, Leipzig: Barth, p. 481–508.

Antonelli, Mauro (1994), *Die experimentelle Analyse des Bewußtseins bei Vittorio Benussi*, Amsterdam / Atlanta: Rodopi (*Studien zur österreichischen Philosophie*, vol. 21).

Antonelli, Mauro (1996), *Percezione e coscienza nell'opera di Vittorio Benussi*, Milan: FrancoAngeli.

Antonelli, Mauro (2005), "Gestalt und Zeit. Historisch-systematische Analysen zur Gestalttheorie", in: J. Ch. Marek and M. Reicher (eds.), *Experience and Analysis. Proceedings of the 27th International Wittgenstein Symposium*, Vienna: Hölder-Pichler-Tempsky, p. 285–295.

Antonelli, Mauro (2006), "Vittorio Benussi: Breve biografia intellettuale – L'attualità di Vittorio Benussi", in: M. Antonelli (ed.), *Vittorio Benussi, Sperimentare l'inconscio. Scritti (1905–1927)*, Milan: Raffaello Cortina, p. 5–120.

Antonelli, Mauro (2018), *Vittorio Benussi in the history of psychology* (*Studies in the history of the philosophy of mind, vol. 21*), Cham: Springer.

Benussi, Vittorio (1902), "Über den Einfluß der Farbe auf die Größe der Zöllner'schen Täuschung", in: *Zeitschrift für Psychologie und Physiologie der Sinnesorgane* 29, p. 264–351; 385–433.

Benussi, Vittorio (1904), "Zur Psychologie des Gestalterfassens (Die Müller-Lyersche Figur)", in: A. Meinong (ed.), *Untersuchungen zur Gegenstandstheorie und Psychologie*, Leipzig: Barth, p. 303–448. Reprinted in V. Benussi, *Psychologische Schriften. Textkritische Ausgabe in 2 Bänden*, Band I. *Psychologische Aufsätze (1904–1914)*, ed. by M. Antonelli, Amsterdam / New York: Rodopi, 2002 (*Studien zur österreichischen Philosophie*, vol. 34), p. 3–138.

Benussi, Vittorio (1905), "La natura delle cosidette illusioni ottico-geometriche", in: S. De Sanctis (ed.), *Atti del V Congresso Internazionale di Psicologia*, Rome: Forzani, p. 262–267. Reprinted in V. Benussi, *Sperimentare l'inconscio. Scritti (1905–1927)*, ed. by M. Antonelli, Milan: Raffaello Cortina, 2006, p. 129–135.

Benussi, Vittorio (1906), "Experimentelles über Vorstellungsinadäquatheit. I. Das Erfassen gestaltmehrdeutiger Komplexe", in: *Zeitschrift für Psychologie* 42, p. 22–55. Reprinted in V. Benussi, *Psychologische Schriften. Textkritische Ausgabe in 2 Bänden*, Band I: *Psychologische Aufsätze (1904–1914)*, ed. by M. Antonelli, Amsterdam / New York: Rodopi, 2002 (*Studien zur österreichischen Philosophie*, vol. 34), p. 155–188.

Benussi, Vittorio (1907a), "Experimentelles über Vorstellungsinadäquatheit. II. Gestaltmehrdeutigkeit und Inadäquatheitsumkehrung", in: *Zeitschrift für Psychologie* 45, p. 188–230. Reprinted in V. Benussi, *Psychologische Schriften. Textkritische Ausgabe in 2 Bänden*, Band I: *Psychologische Aufsätze (1904–1914)*, ed. by M. Antonelli, Amsterdam / New York: Rodopi, 2002 (*Studien zur österreichischen Philosophie*, vol. 34), p. 188–228.

Benussi, Vittorio (1907b), "Zur experimentellen Analyse des Zeitvergleichs. I. Zeitgröße und Betonungsgestalt", in: *Archiv für die gesamte Psychologie* 9, p. 366–449.

Benussi, Vittorio (1908), "Zur experimentellen Analyse des Zeitvergleichs. II. Erwartungszeit und subjektive Zeitgröße", in: *Archiv für die gesamte Psychologie* 13, p. 71–139.

Benussi, Vittorio (1909), "Über 'Aufmerksamkeitsrichtung' beim Raum- und Zeitvergleich", in: *Zeitschrift für Psychologie* 51, p. 73–107. Reprinted in V. Benussi, *Psychologische Schriften. Textkritische Ausgabe in 2 Bänden*, Band I: *Psychologische Aufsätze (1904–1914)*, ed.

by M. Antonelli, Amsterdam / New York: Rodopi, 2002 (*Studien zur österreichischen Philosophie*, vol. 34), p. 229–261.

Benussi, Vittorio (1910), "Über die Grundlagen des Gewichtseindruckes (Beiträge zur Psychologie des Vergleichens)", in: *Archiv für die gesamte Psychologie* 17, p. 1–185.

Benussi, Vittorio (1911), "Über die Motive der Scheinkörperlichkeit bei umkehrbaren Zeichnungen", in: *Archiv für die gesamte Psychologie* 20, p. 363–396. Reprinted in V. Benussi, *Psychologische Schriften. Textkritische Ausgabe in 2 Bänden*, Band I, *Psychologische Aufsätze (1904–1914)*, ed. by M. Antonelli, Amsterdam / New York: Rodopi, 2002 (*Studien zur österreichischen Philosophie*, vol. 34), p. 263–295.

Benussi, Vittorio (1912), "Stroboskopische Scheinbewegungen und geometrisch-optische Gestalttäuschungen", in: *Archiv für die gesamte Psychologie* 24, p. 31–62. Reprinted in V. Benussi, *Psychologische Schriften. Textkritische Ausgabe in 2 Bänden*, Band I: *Psychologische Aufsätze (1904–1914)*, ed. by M. Antonelli, Amsterdam / New York: Rodopi, 2002 (*Studien zur österreichischen Philosophie*, vol. 34), p. 297–328.

Benussi, Vittorio (1913), *Psychologie der Zeitauffassung*, Heidelberg: Winter (*Die Psychologie in Einzeldarstellungen*, ed. by H. Ebbinghaus and E. Meumann, vol. 6). Reprinted in V. Benussi, *Psychologische Schriften. Textkritische Ausgabe in 2 Bänden*, Band II: *Psychologie der Zeitauffassung (1913)*, ed. by M. Antonelli, Amsterdam / New York: Rodopi, 2002 (*Studien zur österreichischen Philosophie*, vol. 35).

Benussi, Vittorio (1914a), "Gesetze der inadäquaten Gestaltauffassung (Die Ergebnisse meiner bisherigen experimentellen Arbeiten zur Analyse der sogen. geometrisch-optischen Täuschungen [Vorstellungen außersinnlicher Provenienz])", in: *Archiv für die gesamte Psychologie* 32, p. 396–419. Reprinted in V. Benussi, *Psychologische Schriften. Textkritische Ausgabe in 2 Bänden*, Band I: *Psychologische Aufsätze (1904–1914)*, ed. by M. Antonelli, Amsterdam / New York: Rodopi, 2002 (*Studien zur österreichischen Philosophie*, vol. 34), p. 341–364.

Benussi, Vittorio (1914b), "Die Atmungssymptome der Lüge", in: *Archiv für die gesamte Psychologie* 31, p. 244–273. Reprinted in V. Benussi, *Psychologische Schriften. Textkritische Ausgabe in 2 Bänden*, Band I: *Psychologische Aufsätze (1904–1914)*, ed. by M. Antonelli, Amsterdam / New York: Rodopi, 2002 (*Studien zur österreichischen Philosophie*, vol. 34), p. 365–394. Eng. transl., "The respiratory symptoms of lying", in: *Polygraph* 4, 1 (1975), p. 52–76.

Benussi, Vittorio (1915), "Monokularlokalisationsdifferenz und haploskopisch erweckte Scheinbewegungen", in: *Archiv für die gesamte Psychologie* 33, Literaturbericht, p. 266–272.

Benussi, Vittorio (1917), "Versuche zur Analyse taktil erweckter Scheinbewegungen (kinematohaptische Erscheinungen) nach ihren äußeren Bedingungen und ihren Beziehungen zu den parallelen optischen Phänomenen", in: *Archiv für die gesamte Psychologie* 36, Literaturbericht, p. 59–135.

Benussi, Vittorio (1918), "Über Scheinbewegungskombination. (Lissajoussche S-, M- und E-Scheinbewegungsfiguren)", in: *Archiv für die gesamte Psychologie* 37, p. 233–282.

Benussi, Vittorio (1922–1923), *Introduzione alla psicologia sperimentale*, Lezioni tenute nell'anno 1922–23 dal Prof. V. Benussi e raccolte dal Dott. C. L. Musatti (Typescript, Benussi Archive, box 7, file 14).

Benussi, Vittorio (1925a), *La suggestione e l'ipnosi come mezzi di analisi psichica reale*, Bologna: Zanichelli. Reprinted in V. Benussi, *Sperimentare l'inconscio. Scritti (1905–1927)*, ed. by M. Antonelli, Milan: Raffaello Cortina, 2006, p. 213–350.

Benussi, Vittorio (1925b), "La suggestione e l'ipnosi come mezzi di analisi psichica reale", in: Società italiana di psicologia (ed.), *Atti del IV Congresso Nazionale di Psicologia (Firenze 1923)*, Florence: Stabilimento Tipografico Bandettini, p. 35–65. Printed also in *Rivista di Psicologia* 21, p. 1–22. Reprinted in V. Benussi, *Sperimentare l'inconscio. Scritti (1905–1927)*, ed. by M. Antonelli, Milan: Raffaello Cortina, 2006, p. 191–211.

Benussi, Vittorio (2002), *Psychologische Schriften. Textkritische Ausgabe in 2 Bänden*, Band I: *Psychologische Aufsätze (1904–1914)*, Band II: *Psychologie der Zeitauffassung (1913)*, ed. by M. Antonelli. Amsterdam / New York: Rodopi (*Studien zur österreichischen Philosophie*, vols 34–35).

Benussi, Vittorio (2006), *Sperimentare l'inconscio. Scritti (1905–1927)*, ed. by M. Antonelli, Milan: Raffaello Cortina.

Benussi, Vittorio / Liel, Wilhelmine (1904), "Die verschobene Schachbrettfigur", in: A. Meinong (ed.), *Untersuchungen zur Gegenstandstheorie und Psychologie*, Leipzig: Barth, p. 449–472.

Biceaga, Victor (2010), *The concept of passivity in Husserl's phenomenology*, Dordrecht: Springer.

Boring, Edwin G. (1929), *A history of experimental psychology*, New York / London: The Century Co.

Boring, Edwin G. (1950), *A history of experimental psychology*, 2nd ed., Englewood Cliffs, NJ: Prentice-Hall.

Bozzi, Paolo (2003), "Anmerkungen zur Praxis und Theorie der experimentellen Phänomenologie. Erinnerungen an meine Erfahrung als Wahrnehmungsforscher in der italienischen gestalttheoretischen Schule", in: *Gestalt Theory* 25, 3, p. 191–198.

Bozzi, Paolo (1980), "Presentazione", in: G. Kanizsa, *Grammatica del vedere. Saggi su percezione e gestalt*, Bologna: Il Mulino, p. 7–22.

Bozzi, Paolo (1985), "La corrente della coscienza, ovvero i fatti sotto osservazione", in: *Teorie & Modelli* 2, 1, p. 5–38. Reprinted in P. Bozzi, *Fenomenologia sperimentale*, Bologna: Il Mulino, 1989, p. 235–273.

Brand, Gerd (1969), *Welt, Ich und Zeit: Nach unveröffentlichten Manuskripten Edmund Husserls*, The Hague: Nijhoff.

Cattaruzza, Serena ed. (1999), *Axiomathes* 10, 1-3, special issue: *Vittorio Benussi.*

Ehrenfels, Christian von (1890), "Ueber 'Gestaltqualitäten'", in: *Vierteljahresschrift für wissenschaftliche Philosophie* 14, p. 249–292. Eng. transl. by B. Smith, "On 'Gestalt qualities'", in: B. Smith (ed.), *Foundation of Gestalt theory*, Munich / Vienna: Philosophia Verlag, p. 82–117.

Held, Klaus (1966), *Lebendige Gegenwart: Die Frage nach der Seinsweise des transzendentalen Ich bei Edmund Husserl, entwickelt am Leitfaden der Zeitproblematik*, The Hague: Nijhoff.

Holenstein, Elmar (1972), *Phänomenologie der Assoziation: Zu Struktur und Funktion eines Grundprinzips der passiven Genesis bei E. Husserl*, The Hague: Nijhoff.

Husserl, Edmund (1900–1901), *Logische Untersuchungen*, 2 vols., vol. 1, ed. by E. Holenstein; vol. 2, in two Parts, ed. by U. Panzer, The Hague: Nijhoff, 1975–1984 (*Husserliana*, XVIII, XIX/1, XIX/2). Eng. transl. by J.N. Findlay, *Logical Investigations*, 2 vols., New York: Humanities Press, 1970.

Husserl, Edmund (1928), *Vorlesungen zur Phänomenologie des inneren Zeitbewußtseins*, ed. by M. Heidegger, in: *Jahrbuch für Philosophie und phänomenologische Forschung* 9, p. 367–498. Reprinted in *Husserliana*, X: *Zur Phänomenologie des inneren Zeitbewußtseins (1893–1917)*, ed. by R. Boehm, The Hague: Nijhoff, 1966. Eng. transl. by J. B. Brough,

On the phenomenology of the consciousness of internal time (1893–1917), The Hague: Kluwer, 1991.

Husserl, Edmund (1948), *Erfahrung und Urteil. Untersuchungen zur Genealogie der Logik*, ed. by L. Landgrebe, Meiner: Hamburg. Eng. transl. by J. Churchill and K. Ameriks, *Experience and judgment*, Evanston, IL: Northwestern University Press, 1973.

Husserl, Edmund (1966), *Analysen zur passiven Synthesis. Aus Vorlesungs- und Foschungsmanuskripten (1918–1926)*, ed. by M. Fleischer, The Hague: Nijhoff (*Husserliana*, XI). Eng. transl. by A. J. Steinbock, *Analyses concerning passive and active synthesis: Lectures on transcendental logic*, Dordrecht: Kluwer, 2001.

Husserl, Edmund (1973), *Zur Phänomenologie der Intersubjektivität: Texte aus dem Nachlass. Dritter Teil: 1929-1935*, ed. by I. Kern, The Hague: Nijhoff, 1973 (*Husserliana*, XV).

Kanizsa, Gaetano (1952), "Legittimità di una analisi del processo percettivo fondata su una distinzione in 'fasi' o 'stadi'", in: *Archivio di Psicologia, Neurologia e Psichiatria* 13, p. 292–322.

Kanizsa, Gaetano (1976), "Subjective Countours", in: *Scientific American* 234, 4, p. 48–52.

Kanizsa, Gaetano (1979), *Organization in vision*, New York: Praeger.

Kanizsa, Gaetano (1980), *Grammatica del vedere*, Bologna: Il Mulino.

Kanizsa, Gaetano (1991), *Vedere e pensare*, Bologna: Il Mulino.

Kern, Iso (1964), *Husserl und Kant: Eine Untersuchung über Husserls Verhältnis zu Kant und zum Neukantianismus*, The Hague: Nijhoff.

Koffka, Kurt (1915), "Beiträge zur Psychologie der Gestalt- und Bewegungserlebnisse. III. Zur Grundlegung der Wahrnehmungspsychologie. Eine Auseinandersetzung mit V. Benussi", in: *Zeitschrift für Psychologie* 73, p. 11–90.

Koffka, Kurt (1935), *Principles of Gestalt psychology*, London: Kegan Paul, Trench, Trubner & Co.

Köhler, Wolfgang (1920), *Die physischen Gestalten in Ruhe und im stationären Zustand*, Braunschweig: Vieweg.

Köhler, Wolfgang (1947), *Gestalt Psychology. An introduction to new concepts in modern psychology*, New York: Liveright.

Kühn, Rolf (1998), *Husserls Begriff der Passivität: Zur Kritik der passiven Synthesis in der genetischen Phänomenologie*, Freiburg i. Br. / Munich: Alber.

Linke, Paul Ferdinand (1918), *Grundfragen der Wahrnehmungslehre. Untersuchungen über die Bedeutung der Gegenstandstheorie und Phänomenologie für die experimentelle Psychologie*, München: Reinhardt.

Meinong, Aexius (1889), "Über Phantasie-Vorstellung und Phantasie", in: *Zeitschrift für Philosophie und philosophische Kritik* 95, p. 161–244. Reprinted in *Gesamtausgabe*, I, p. 109–192.

Meinong, Aexius (1894), "Beiträge zur Theorie der psychischen Analyse", in: *Zeitschrift für Psychologie und Physiologie der Sinnesorgane* 6, p. 340–385, 417–455. Reprinted in *Gesamtausgabe*, I, 305–388. Eng. transl. by M.-L. Schubert Kalsi, "An essay concerning the theory of psychic analysis", in: A. Meinong, *On objects of higher order and Husserl's phenomenology*, ed. by M.-L. Schubert Kalsi, The Hague / Boston / London: Nijhoff, p. 73–135.

Meinong, Aexius (1902), *Über Annahmen*, Leipzig: Barth (*Zeitschrift für Psychologie und Physiologie der Sinnesorgane*, Ergänzungsband 2). Partly reprinted in *Gesamtausgabe*, IV, p. 387–489.

Meinong, Aexius (1910), *Über Annahmen*, zweite, umgearbeitete Auflage, Leipzig: Barth. Reprinted in *Gesamtausgabe*, IV, p. 1–389, 517–535. Eng. transl. by J. Heanue, *On Assumptions*, ed. by J. Heanue, Berkeley / Los Angeles, CA: University of California Press.

Messer, August (1908), *Empfindung und Denken*, Leipzig: Quelle & Meyer.

Metzger, Wolfgang (1954), *Psychologie. Die Entwicklung ihrer Grundannahmen seit der Einführung des Experiments*, 2nd ed., Darmstadt: Steinkopff.

Miller, Izchak (1984), *Husserl, perception, and temporal awareness*, Cambridge, MA: The MIT Press.

Musatti, Cesare (1928), "L'attività scientifica di Vittorio Benussi", in: *Archivio Italiano di Psicologia* 6, p. 259–273.

Rang, Bernhard (1990), *Husserls Phänomenologie der materiellen Natur*, Frankfurt a. M.: Klostermann.

Römpp, Georg (1992), *Husserls Phänomenologie der Intersubjektivität und ihre Bedeutung für eine Theorie intersubjektiver Objektivität und die Konzeption einer phänomenologischen Philosophie*, Dordrecht: Kluwer.

Rubin, Edgar (1915a), "Die visuelle Wahrnehmung von Figuren", in: F. Schumann (ed.), *Bericht über den VI. Kongreß für experimentelle Psychologie in Göttingen vom 15. bis 18. April 1914*, Leipzig: Barth, p. 60–63.

Rubin, Edgar (1915b), *Synsoplevede figurer. Studier i psykologisk analyse,* København og Kristiania: Gyldendalske, Nordisk forlag. Ger. transl., *Visuell wahrgenommene Figuren: Studien in psychologischer Analyse*, Copenhagen: Gyldendal, 1921.

Sandmeyer, Bob (2009), *Husserl's constitutive phenomenology: Its problem and promise*, London: Routledge.

Schuhmann, Karl (1975), *Index nominum zum Nachlaß von Edmund Husserl*, Leuven.

Sinatra, M. (1998), "Vittorio Benussi", in: G. Cimino and N. Dazzi (eds.), *La psicologia in Italia. I protagonisti ei problemi scientifici, filosofici e istituzionali (1870–1945)*, Milan: LED, p. 473–521.

Smith, Barry ed. (1988a), *Foundation of Gestalt theory*, Munich / Vienna: Philosophia Verlag.

Stern, L. William (1897), "Psychische Präsenzzeit", in: *Zeitschrift für Psychologie und Physiologie der Sinnesorgane* 13, p. 325–349. Eng. transl. by N. de Warren, "Mental presence-time", in *The New Yearbook for Phenomenology and Phenomenological Philosophy* 5, 1, 2005, p. 325–351.

Taguchi, Shigeru (2006), *Das Problem des "Ur-Ich" bei Edmund Husserl: Die Frage nach der selbstverständlichen "Nähe" des Selbst*, Dordrecht: Springer.

Wertheimer, Max (1923), "Untersuchungen zur Lehre von der Gestalt. II.", in: *Psychologische Forschung* 4, p. 301–350. Eng. transl. by M. Wertheimer and K. W. Watkins, "Investigations on Gestalt principles", in: M. Wertheimer, *On perceived motion and figural organization*, ed. by L. Spillmann and M. Wertheimer, Cambridge, MA: The MIT Press, p. 127–182.

Witasek, Stephan (1908), *Grundlinien der Psychologie*, Leipzig: Dürr.

Yamaguchi, Ichiro (1982), *Passive Synthesis und Intersubjektivität bei Edmund Husserl*, The Hague: Nijhoff.

Archival Material

Benussi Archive (Archivio Storico della Psicologia Italiana, University of Milano-Bicocca).

Index of Names

Albertazzi, Liliana 15, 26, 30, 70ff., 143
Alpi, Ilaria 156
Ameseder, Rudolf 1, 26f., 30, 47, 53, 56ff., 69, 74, 171, 174
Antonelli, Mauro 7, 29f., 72f., 75ff., 142, 169f.
Aristotle 5, 91, 94, 101, 147, 154

Barber, Kenneth 17f., 30
Barceló, Axel A. 38
Baumgartner, Wilhelm 126
Bayne, Tim 85
Bealer, George 81
Bengson, John 79
Benussi, Vittorio 1, 6, 15, 20, 27ff., 47, 53, 55, 58, 61ff., 67ff., 74ff., 169ff., 178ff., 184ff.
Benussi-Liel, Wilhelmine 20, 27, 173
Berger, Harald 80
Berghofer, Philipp 61, 68
Bergmann, Gustav 164
Berkeley, George 31, 107, 123, 142f., 158, 163, 197
Biceaga, Victor 183
Boccaccini, Federico 106
Böhm-Bawerk, Eugen von 125
Boltzmann, Ludwig 23
Bolzano, Bernard 1, 158
Boring, Edwin 15, 30, 75, 169f., 193
Boyd, Kenneth 65f.
Bozzi, Paolo 178f., 189
Bradley, Francis Herbert 22
Brand, Gerd 183
Brandt, Richard 70
Brenner, William H. 42
Brentano, Franz 1ff., 15ff., 24, 30, 32, 47, 69f., 73, 76f., 82f., 85ff., 88ff., 98ff., 112f., 126, 128, 146, 152, 161ff., 179f.
Breyer, Thiemo 85
Broad, Charlie 103f.
Bühler, Karl 80
Buridan, Johannes 63

Cappelen, Hermann 65
Carvalho, Marcelo 38
Castañeda, Héctor-Neri 2, 168
Cattaruzza, Serena 170
Centi, Beatrice 141
Chrudzimski, Arkadiuz 15, 24, 30, 126
Chudnoff, Elijah 79
Cokely, Edward 66
Crone, Robert A. 35
Cucchi, Stefano 156
Cullen, Simon 66f.

De Sousa, Ronald 145f.
Deonna, Julien A. 114, 118
Descartes, René 22, 101, 103, 105, 162f.
Deutsch, Max 65
Dewalque, Arnaud 1, 85ff., 89, 93
Dölling, Evelyn 3, 6f., 71, 74
Donato-Rodriguez, Xavier 138
Dos Santos, Miguel F. 122
Dummett, Michael 43

Eaton, Howard O. 126
Ebbinghaus, Hermann 35, 195
Eco, Umberto 109f.
Ehrenfels, Christian von 18f., 32, 47ff., 58f., 115ff., 127, 143, 174f., 196
Erhard, Christopher 78

Faist, Anton 22
Fechner, Gustav Theodor 21
Feltz, Adam 66
Findlay, John N. 8, 125, 130, 158, 196
Frankl, Wilhelm Maria 27
Frege, Gottlob 100

Geiger, Moritz 119
Gettier, Edmund 62f., 65f., 82f.
Ghijsen, Harmen 78f.
Gödel, Kurt 67
Goldman, Alvin 64
Goodman, Nelson 43
Graff, Delia 43
Grossmann, Reinhard 9, 14, 30, 158, 161, 164, 167f.

https://doi.org/10.1515/9783110664850-012

Guigon, Ghislain 21, 30, 71
Gutland, Christoher 85

Haller, Rudolf 8, 13, 17f., 21, 30, 71, 123, 142, 158
Hardin, Clyde Laurence 43f.
Heanue, James 107, 123, 141, 143, 158, 197
Held, Klaus 183
Helmholtz, Hermann 35
Hillebrand, Franz 4
Hintikka, Jaalo 2
Höfer, Ulf 7, 71, 80
Höfler, Alois 3ff., 11ff., 16ff., 20f., 23f., 28, 30f., 35, 38, 94, 96, 180
Holenstein, Elmar 183
Horgan, Terence 107
Hrovatin, Miran 156
Huber, Helmuth P. 15, 29, 31
Huemer, Michael 78, 123, 142f.
Husserl, Edmund 1, 19, 30, 47., 61, 69ff., 75f., 78ff., 86, 93, 128, 179ff., 186

Ichikawa, Jonathan 65
Ierna, Carlo 47f., 53

Jackson, Frank Cameron 43, 46
Jacquette, Dale 2, 15, 24, 30f., 138

Kanizsa, Gaetano 178f., 193
Kant, Immanuel 31, 162ff., 197
Kauppinen, Antti 66
Kenny, Anthony 119
Kern, Iso 183
Kindinger, Rudolf 8, 13, 17, 30, 142, 158
Knobe, Jushua 65f., 76, 83
Koffka, Kurt 170, 175ff., 179
Köhler, Wolfgang 176ff.
Kriegel, Uriah 86, 88, 90, 93, 103ff.
Kuehni, Rolf G. 35
Kühn, Kolf 183
Külpe, Oswald 107, 119

Ladd, John 70
Lambert, Karel 2, 46
Lamedica, Edoardo 106
Lange, Carl 119
Lauria, Federico 114
Lenggenhager, Bigna 68
Liel, Wilhelmine cf. Benussi-Liel, Wilhelmine
Lindenfeld, David 15, 23, 31, 125, 132
Linke, Paul Ferdinand 185
Lipps, Theodor 1
Locke, John 70f., 101, 105

Mach, Ernst 47
Machery, Edouard 66
Madary, Michael 80
Maier, Anneliese 63
Mally, Ernst 1, 18, 27, 47, 53ff., 58
Manotta, Marina 72f., 77, 79
Marek, Johann Ch. 29, 31, 69, 74, 80, 132, 194
Martinak, Eduard 6ff.
Martinelli, Riccardo 11, 21, 31
Marty, Anton 87, 94, 100f., 142
Massin, Oliver 114
Meinong, Alexius 1ff., 11ff., 42ff., 56, 58ff., 65, 69ff., 84ff., 89, 92, 94ff., 125ff., 145ff., 161f., 164ff., 170f., 174f., 179f., 184, 194
Meister, Richard 32
Menger, Carl 125
Messer, August 186
Metzger, Wolfgang 177f., 198
Metzinger, Thomas 68
Miller, Izchak 183
Montague, Michelle 85
Mortensen, Kaija 67
Mulligan, Kevin 34, 38, 43, 47, 119, 145f., 155
Münch, Dieter 73, 80
Musatti, Cesare 170, 178, 195

Nadelhoffer, Thomas 63f.
Nado, Jennifer 64
Nagel, Jennifer 65ff.
Nahmias, Eddy 63f.
Nicholas of Autrecourt 63
Nichols, Shaun 66

Ostwald, Wilhelm 35, 38

Park, Byong-Chul 38
Papineau, David 43, 46
Parsons, Terence 2

Paśniczek, Jacek 2
Pinillos, Ángel 66
Pinkerton, R.J. 43, 46
Plato 162
Poli, Roberto 15, 30, 81, 143
Prinz, Jesse 67
Pust, Joel 64

Rang, Bernhard 183
Rapaport, William J. 2
Raspa, Venanzio 1, 8f., 70, 106, 110, 125, 127, 139, 145ff., 149f., 154f.
Regeni, Giulio 156
Reicher, Maria E. 117, 125, 194
Reinach, Adolf 93, 102
Rescher, Nicolas 43
Richard, Sébastien 100ff.
Rickert, Heinrich 2f.
Rinofner-Kreidl, Sonja 68f., 80
Rollinger, Robin 37f., 106
Röck, Tina, 87
Römpp, Georg 183
Routley, Richard 2
Rubin, Edgar 172
Runge, Philipp Otto 35
Russell, Bertrand 1, 11, 22, 32, 71, 98f.
Ryle, Gilbert 119

Saab, Salma 38
Sandmeyer, Bob 183
Saxinger, Robert 27
Schubert Kalsi, Marie-Luise 8, 13, 17, 30, 32, 143, 158, 197
Schuhmann, Karl 107, 126f., 181
Schwarz, Andreas 35
Seron, Denis 33, 35, 106
Seyedsayamdost, Hamid 66
Siegel, Johanna 78
Sinatra, Maria 170
Smith, Barry 24, 32, 47, 86, 125, 196
Stern, L. William 187
Stich, Stephen 70
Stout, George 89, 92, 103
Stumpf, Carl 5, 22f., 31f., 47
Sylvan, Richard cf. Routley, Richard

Taguchi, Shigeru 183
Tappolet, Christine 114, 123, 151
Tegtmeier, Erwin 21, 32, 161f.
Teroni, Fabrice 114, 118
Tienson, John 87
Tobia, Kevin 70
Turri, John 66
Twardowski, Kazimierz 1, 3f., 8f., 23f., 32, 87, 103f., 158

Urban, Wilbur 27f., 32, 126

Vendrell Ferran, Íngrid 76, 119, 122, 129, 139

Wallner, Michael 80
Walton, Kendall 119, 121f.
Weinberg, Jonathan 64ff.
Wertheimer, Max 176
Wieser, Friedrich von 125
Williamson, Timothy 64f.
Wiltsche, Harald 80
Windt, Jennifer 68
Witasek, Stephan 1, 15, 22f., 28f., 31f., 69, 74, 96, 99, 119, 171, 180
Wittgenstein, Ludwig 33, 35, 38, 40, 43ff., 142, 194
Wright, Crispin 43
Wright, Jennifer 65
Wundt, Wilhelm 21, 35, 73, 190

Yamaguchi, Ichiro 183, 198

Zalta, Edward 2, 31

Bei Fragen zur Produktsicherheit wenden Sie sich bitte an:
If you have any questions regarding product safety,
please contact:

Walter de Gruyter GmbH
Genthiner Straße 13
10785 Berlin
productsafety@degruyterbrill.com